# Instrumentation and Control Systems Documentation

# Instrumentation and Control Systems Documentation

Frederick A. Meier and Clifford A. Meier

ISA—The Instrumentation, Systems, and Automation Society

**Notice**

The information presented in this publication is for the general education of the reader. Because neither the author nor the publisher have any control over the use of the information by the reader, both the author and the publisher disclaim any and all liability of any kind arising out of such use. The reader is expected to exercise sound professional judgment in using any of the information presented in a particular application.

Additionally, neither the author nor the publisher have investigated or considered the affect of any patents on the ability of the reader to use any of the information in a particular application. The reader is responsible for reviewing any possible patents that may affect any particular use of the information presented.

Any references to commercial products in the work are cited as examples only. Neither the author nor the publisher endorses any referenced commercial product. Any trademarks or trade names referenced belong to the respective owner of the mark or name. Neither the author nor the publisher makes any representation regarding the availability of any referenced commercial product at any time. The manufacturer's instructions on use of any commercial product must be followed at all times, even if in conflict with the information in this publication.

Copyright © 2004 ISA – The Instrumentation, Systems, and Automation Society

All rights reserved.

Printed in the United States of America.
10 9 8 7 6 5 4 3

ISBN 1-55617-870-0

No part of this work may be reproduced, stored in a retrieval system, or transmitted in any form or by any means, electronic, mechanical, photocopying, recording or otherwise, without the prior written permission of the publisher.

ISA 67 Alexander Drive
P.O. Box 12277 Research Triangle Park, NC 27709

**Library of Congress Cataloging-in-Publication Data**

Meier, Fred A.
  Instrumentation and control systems documentation / Fred A. Meier and Clifford A. Meier.
     p. cm.
  ISBN 1-55617-870-0
  1. Process control. 2. Engineering instruments. 3. Technology–Documentation. I. Meier, Clifford A. II. Title.
  TS156.8.M45 2004
  670.42'7–dc22
                    2004011286

## DEDICATION

*This book is dedicated to Cris and Jean,*
*Without whose assistance and encouragement*
*this book would not have been started,*
*let alone finished.*

## ACKNOWLEDGMENTS

We wish to thank all those who assisted in the development of this book, especially
Dave Fusaro of *Control*, a Putman Media Co. publication;
Ken Brabham of Industrial Design Corporation; co-workers at Harris Group Inc.;
and to the Technical and Education Services Departments of ISA, especially Lois Ferson, Alice Heaney and
Linda Wolffe; designer, Vanessa F. Harris;  and our copyeditor, Jim Strothman.

Cliff would like to especially thank the lads in Dublin, for making it fun, and in particular to Tony Riordan,
whose encouragement to have that ceremonial first scoop with "me Da" eventually
led to the idea of writing this book.

# Contents

# List of Illustrations

## ABOUT FRED MEIER

Fred Meier's career spans more than 50 years as a control systems engineer, chief engineer, and engineering manager in the oil, chemical and engineering industries in the United States, Algeria, Canada, Germany, Japan, and the United Kingdom. He has held Professional Engineer licenses in New York, New Jersey, California, Alberta, Manitoba, and Saskatchewan. He completed U.S. Army training as an electrical engineer and has a Mechanical Engineering Degree from Stevens Institute of Technology and an MBA from Rutgers University.

Fred has been an ISA member more than 40 years. He has served as President of the New York Section; the Edmonton, Alberta, Section; and the Tarheel (North Carolina) Capital Area Section. He was awarded the ISA District II Golden Eagle Award in 2000.

Fred presented two papers at ISA 1982, "Why Not Be An Adaptive Manager?" and, jointly with co-worker Trevor Haines, "Contractor Handling of Engineering for Distributed Control Systems". He authored the cover article for *CHEMICAL ENGINEERING*, Feb. 22, 1982, "Is your control system ready to start up?". Fred and son Cliff (this book's co-author) presented a joint paper at ISA 1999, "A Standard P&ID, Elusive as the Scarlet Pimpernel". Fred also published an editorial viewpoint in *ISA TRANSACTIONS*, October 2002, "A P&ID standard: What, Why, How?".

After Fred's "first" retirement, he served as the ISA Staff Engineer; after his "second" retirement, as an ISA Instructor and Consultant; and, since his "third" retirement, as co-author of this book. Fred and Jean have been married for 56 years, and are the proud parents of four children, four grandchildren, and one great granddaughter. They currently live in Chapel Hill, North Carolina.

## ABOUT CLIFF MEIER

Cliff Meier's 26 years of engineering experience started with a Bachelor of Science degree in Mechanical Engineering from Northeastern University. His attraction to the widgets and intricacies of Instrumentation and Controls has taken him to three continents and to industrial controls projects in nuclear and fossil fuel power generation, oil and gas production, chemical and pulp and paper industries, and microelectronics factories. Cliff has worked exclusively in consulting engineering on projects ranging in complexity from a few loops to complex modernization projects and greenfield installations entailing thousands of loops. His career started with manual drafting on Mylar sheets and has transitioned to computer-aided design (CAD), where data handling has almost eclipsed the importance of the physical drawings. While he enjoys the team relationships of industrial design projects, he finds construction and commissioning work to be almost as rewarding as writing with his Dad.

Cliff is a member of ISA and holds professional engineering licenses in Texas and Oregon.

Cliff and his wife, Cris, have been married more than 25 years and are the parents of the two finest kids on earth — Will and Helen. They live in Beaverton, Oregon, where they can gaze at snow-capped mountains when it isn't raining, which, come to think of it, is not that often.

# Instrumentation and Control Systems Documentation

## Introduction

There are three types of processes in industry: continuous, batch, and discrete manufacturing. A brief description of each type follows:

**Continuous.** Material is fed into and removed from the process at the same time. Petroleum refining is a good example.

**Batch.** Material is added to a vessel or other piece of equipment, some process takes place, and the changed material is subject to another step. Many repeats of the above steps, perhaps using different equipment, may be necessary to make the finished product. Beer, for example, is made by a batch process.

**Discrete manufacturing.** Separate components, parts or sub-assemblies are manufactured or assembled to produce a product. Automobile manufacturing is an example.

The process industry sector of the worldwide economy consists of plants that operate continuously and those that operate in batch mode. Since there are similarities in design and operation, plants that operate continuously and those that operate in batch mode are generally combined under the "process industries" label. All documents discussed in this book are common in process industries.

The nature of the documentation we use to describe modern instrumentation and control systems has evolved over many years to maintain a primary objective – to impart efficiently and clearly salient points about a specific process to the trained viewer. As processes we are concerned with become more complex, so then does the documentation. An ancient simple batch process like making brine might be defined quite clearly without so much as a schematic drawing, simply by showing a few pipes, a tank and some manual valves. A modern continuous process that runs twenty-four hours a day, seven days a week, with specific piping and valve requirements, many interrelated controls, numerous monitoring points, operator control requirements, pumps, motorized equipment and safety systems will, of course, require a more complex documentation system. Figure I-1 shows examples of typical continuous processes.

| **Figure I-1: Typical Continuous Processes** |
|---|
| • Steam production |
| • Chemical reactions |
| • Separations |
| • Waste treatment |
| • Distillation |

As the amount of information needed to define the process increases, the documents must become more specialized, allowing for efficient grouping of details. The piping design group develops and maintains their line lists; the instrumentation and controls design group does the same with their Instrument Lists. Although both lists are keyed to a general document in some simple way, the lists themselves are extremely detailed and lengthy, containing information of value to specialists but not necessarily important to others.

General information that defines a process is maintained in a form that is both simple and easily read, but without all of the detailed information needed by a specialist. An example of a general document is a Piping and Instrumentation Drawing (P&ID). The general document serves as the key to the more detailed documents. Information presentation and storage has then become more efficient. The overall picture and shared information of use to most people are on the general document. Information of use to specialists to flesh out the design is maintained on the detailed document.

The documents that describe modern industrial processes, like most technical work, assume some level of understanding on the reader's part. The documents use a schematic, symbol-based "language" that may resemble Mayan hieroglyphics to those unfamiliar with the process nomenclature. The symbols, however, include a wealth of information to those trained to translate them. Both tradition and standards also govern the presentation of these symbols on the document. Indeed, the very existence of some types of documents may seem odd unless the observer understands their intended function. Like any live modern language, the symbols and their applications are being improved constantly to meet new challenges.

This book will train you to read, understand, and apply the symbols and documents used to define a modern industrial instrumentation and control system. For more experienced professionals, we will offer insights into using the symbols and documents effectively, including explanations for their use. We will present variations in the use of symbols and documents we have seen, and point out some pitfalls to avoid. To better understand process design documentation today, we will look at how and when documents are developed, who develops them, why they are developed, and how they are used. The types of documents we will discuss include Process Flow Diagrams, Piping and Instrumentation Drawings, Instrument Lists, Specification Forms, Logic Diagrams, Installation Details, Location Plans, and Loop Diagrams. We also will investigate how these documents can be used to best advantage during plant construction and operation.

The authors are strong proponents of honoring and using standards, including industry standards developed by ISA — The Instrumentation, Systems, and Automation Society and other organizations, and plant standards developed especially for a specific location. However, we are not zealots. The documentation must fulfill a need and not present information simply because you perceive it is called for by some standard. That said, you should understand standards are almost always much more "experienced" than you are. They have been developed, reviewed, and time tested by people from every industry and every function within those industries. You should not deviate from standards unless you have carefully considered all the ramifications of doing so. For example, we know of one company that does not use Loop Diagrams. They have been able to meet their maintenance, configuration, construction, and purchasing requirements with some very creative use of instrument databases. However, they arrived at the stage where they felt confident changing their documentation "set" after carefully considering and testing some assumptions. They reviewed the proposed document set with all concerned parties, including their design and construction contractors and their own management before committing to using databases in lieu of Loop Diagrams.

Process documents have to "work" to be effective. Plant design and operations personnel using them must have confidence the information shown is accurate and up-to-date. A facility might be operating unsafely if there is no culture or system in place for updating documents. If this pipe no longer connects to that piece of equipment, is that associated relief valve still protecting what it should? Can the controls still maintain temperature if there is insufficient coolant due to undocumented tie-ins that have depleted the available cooling water? If documents are not up-to-date, future changes to your facility will be extraordinarily and needlessly expensive. Any reputable contractor will verify the current condition of the process before implementing changes. An effective change must be made based upon *what you have* rather than *what you had*.

The modern industrial facility can be chaotic at times. However, plant and project personnel must be able to communicate easily. An industry-recognized language will facilitate that communication. Design projects are difficult enough in today's economic environment without the additional work-hour burden of developing unique symbols to define systems when a more recognized and understood system is already available. And, believe us, someone in your design firm right now is probably doing just that.

The industry standards we discuss in this book have been tested over time, and they work. We will explain how and why they work; it is up to you to apply this knowledge. Of course, documentation you use and its content must stand the "customer" test. They must be of value to the customer, and they must be useful! A perfectly executed Loop Diagram with all the features outlined in ISA-5.4-1991, Instrument Loop Diagrams is of little value if no one uses it. We also want to point out that industry standards allow you to make variations in the content of the documentation to suit your specific requirements.

This book will be easy to read, with many illustrations and little or no mathematics (and absolutely no calculus!). It will be of interest to engineers and technicians, not only in the instrumentation and control field, but in many other specialties as well. Instrumentation and controls design groups are unique in that they have to coordinate among all the other disciplines in a plant, mill, or factory during design and operation. This book will explain their varied, encompassing language. It will also be of value to plant operating, maintenance, and support personnel who are interested in plant design "deliverables" – the documentation that an instrumentation and controls design group usually develops.

The engineering design phase of a typical continuous process plant may last from perhaps a few months to several years. Once the plant is built it may operate for thirty or more years. Common sense dictates that the documents developed during the engineering phase should have lasting value throughout a plant's operating life.

The purpose of the book is to provide the reader with enough information to be able to understand the documents and the information on them and to use that understanding effectively. It is hoped this knowledge will be useful, not only in existing plants, but also as a basis for a review and reality check on future engineering design packages. Also — dare we say it — we hope to encourage effective discussions among the design team, the construction contractor, and the maintenance team that will lead them to agree on the documentation set that will most effectively meet all their requirements.

The documents we will look at in this book have been developed over time to very efficiently meet the needs of plant construction. As personnel with more technical expertise scrutinize these documents closely, perhaps they can improve them.

We will look at instrumentation and controls documents in two ways. First, we will look with enough detail to help the reader understand the form and function; then we will review the application. For some of these documents, no published industry standard is available to guide us about their content. We will therefore describe what we believe is a middle path — one many will accept but, realistically, one that may not be accepted by everyone in every detail.

You may have heard standards developers say "my way or the highway" or "there are two ways to do anything, my way and the wrong way". They take this approach from necessity, since a wishy-washy industry standard is not much of a "standard"; it has little value. We will not be as dogmatic, since we want you to develop a documentation set that works for your facility – one that meets your specific requirements. We believe it is appropriate to develop plant documentation standards for your facility democratically – with input from all the parties that have a stake in the product, as well as ones that honor the industry standards. However, we urge you to control changes to that plant standard very carefully once a majority of your users have defined your documentation requirements. Rigid control is critical for an effective system. Develop freely; operate rigidly.

The second way we will look at a typical document set is to use a very simple simulated project to follow the sequence by which the documents are developed. There is a logical sequence in their preparation. Often one document type must be essentially complete before the next type of document can be started. If the documents are not developed in the right sequence, work-hours will be wasted, since you will have to revisit the document later to incorporate missing information. While the sequence is of more importance to those interested in the design process, it is useful for operating personnel to understand how document sets are developed. If done for no other reason, this understanding will help ensure operating personnel modify all the information in all the affected documents as they make changes.

In our experience, there are many different ways to define instrumentation and control systems. All the plants that used the markedly different document sets were, eventually, built and operated. Of course, some projects ran smoothly, while others seemed to develop a crisis a minute. Some were easier to build, and some took longer, but eventually all the plants were completed. Sometimes, the document set's content had a direct influence on how well the project ran, and a smoothly run project is a less expensive project.

The use of computers in engineering design is offering new ways to define the work to be performed. Indeed, the new ways available now with linked documents offer attractive efficiency and accuracy that may compel some to revisit the content of the standard design package document set.

The eight document types listed below — discussed in detail in this book — have been used successfully as a "set" of documents for many years.

## Process Flow Diagram

The Process Flow Diagram (PFD) defines the process schematically. It shows what and how much of each product the plant will make; quantities and types of raw materials necessary to make the products; what by-products are produced; the critical process conditions — pressures, temperatures, and flows necessary to make the product; and major piping and equipment necessary. For a very simple PFD, see Chapter 1, Figure 1-1 (page 12).

### Piping and Instrumentation Drawing

The Piping and Instrumentation Drawing (P&ID) is the overall design document for a process plant. It defines – using symbols and word descriptions – the equipment, piping, and the instrumentation and control system. It is also the key to other documents. For example instrument tag numbers are shown on a P&ID. This tag number is the key to finding additional information about this device on many other documents. The same is true for line and equipment numbers. For a P&ID, see Chapter 2, Figure 2-21 (page 45).

### Instrument List

The Instrument List is an alphanumeric list of the data related to a facility's instrumentation and control systems components and, possibly, functions. Instrument Lists are organized using the alphanumeric tag numbers of the instrumentation and control system devices. They reference the various documents that contain the information needed to define the total installation. Instrument Lists are discussed in Chapter 3.

### Specification Forms

The Specification Forms or instrument data sheets define each tag-numbered instrumentation and control system device with sufficient detail so a supplier can quote and eventually furnish the device. For a typical Specification Form, see Chapter 4, Figures 4-4, 4-5 and 4-6 (pages 74-76).

### Logic Diagrams

Logic Diagrams are the drawings used to design and define the on-off or sequential part of a continuous process plant. For a typical Logic Diagram, see Chapter 6, Figure 6-5 (page 101).

### Loop Diagrams

A Loop Diagram is a schematic representation of a single control loop (sensing element, control component, and final element). It depicts the process connections and the components' interconnection to the power sources and transmission systems (pneumatic, electronic, or digital). For a typical Loop Diagram, see Chapter 7, Figure 7-1 (page 108).

### Installation Details

Installation Details are used to show how the instrumentation and control system components are connected and interconnected to the process. They provide the methods the plant uses to support the devices and the specific requirements for properly connecting to the process. Installation Details are discussed in Chapter 8.

## Location Plans

Location Plans are orthographic views of the plant, drawn to scale, that show the locations of instruments and control system components. They often show other control system hardware including marshaling panels, termination racks, local control panels, junction boxes, instrument racks, and perhaps power panels and motor control centers. Location Plans are discussed in Chapter 8.

These eight document types are developed sequentially as the project progresses and as the relevant information becomes available. See Figure I-2, Instrument Drawing Schedule, which illustrates typical sequential document development.

The Process Flow Diagram (PFD) is the starting point for designing any process plant. It is the macroscopic, schematic view of the major features of a process or facility; it is the "talking document" for managers, planners and the process design team. The instrumentation and controls design group has little involvement in developing the PFD, due to its macroscopic nature. The PFDs are used to develop a project scope; they may also be used (and maintained) to

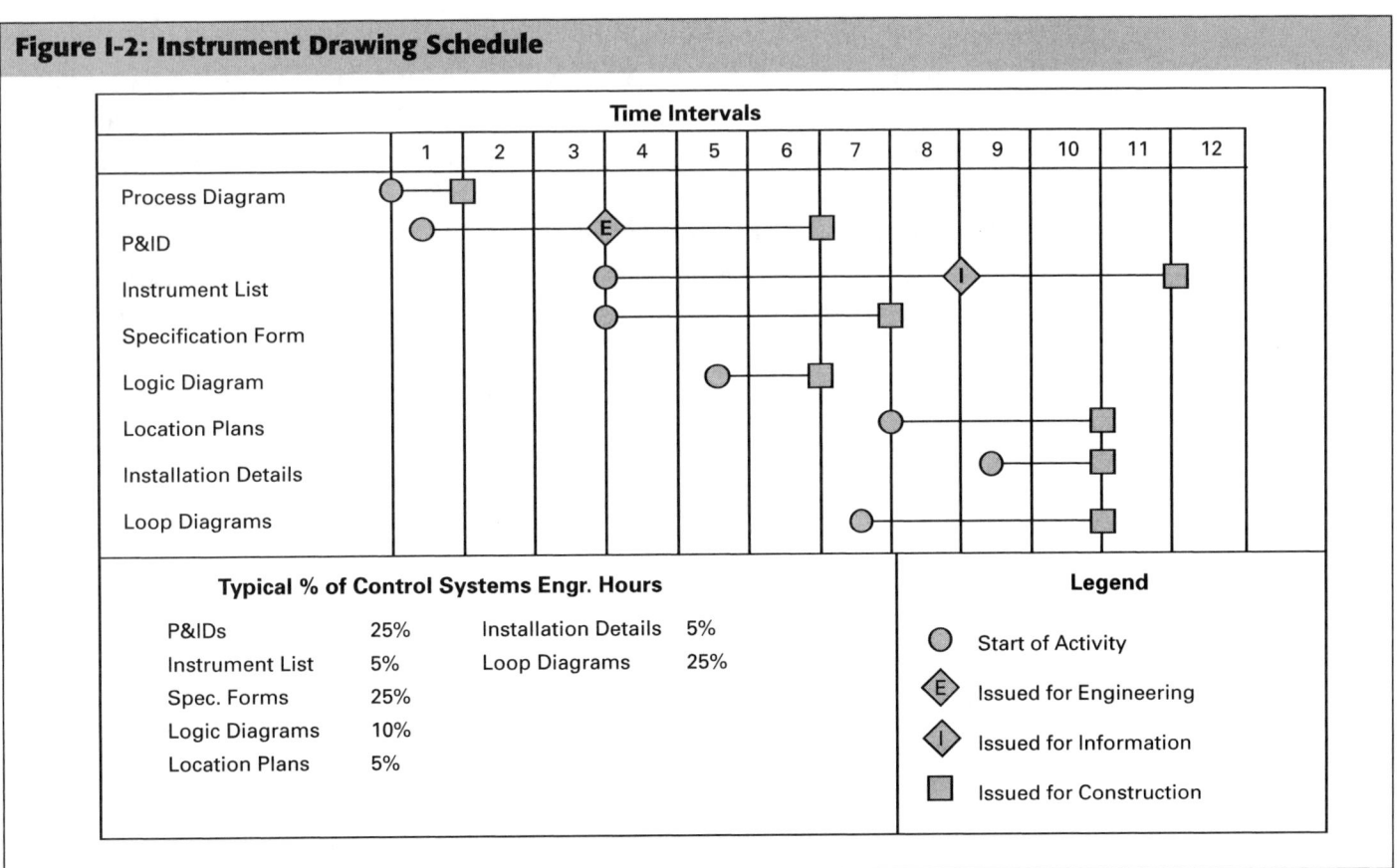

**Figure I-2: Instrument Drawing Schedule**

document overall product and utility balances. For any specific project, PFDs are normally issued for the purpose of gathering comment and review. After questions and clarifications are resolved, the general scope is essentially established, and the P&IDs are then started along with detailed scoping, estimating and design processes.

Developing P&IDs is a very interactive process. Specialists designing vessels, mechanical equipment, piping, process, electrical, instrumentation and control systems all provide input into their development. Each specialist group puts information on the drawing in a standardized way, adding details as they become available.

We will discuss symbols and tag numbers in greater detail in Chapter 2. Briefly, a symbol defines the type of instrument, and the tag number uniquely identifies that specific instrument. A tag number consists of a few letters that describe the device, plus a combination of number and letters to uniquely identify it. See Figure I-3 for an example of an instrument that might be indicated on a P&ID. The circle shows a field-mounted instrument located on a pipe. The "PI" further describes the device as a pressure indicator or gauge. In this instance, sequential numbering is used. Since the gauge is the first of its type on the P&ID, the "loop" number "1" is used. The next pressure gauge in this numbering system would have the tag number "PI-2".

**Figure I-3: Field Mounted Pressure Gauge**

PI
1

Some tag numbers are much more complex. See Figure I-4 for a very complex tag number: "1A-AA-PT-100-A". "1A-AA" is a prefix that designates a unit definition (i.e., a part or unit of the plant) called "1A" and a system designator (i.e., a process system within the plant unit) called "AA". The "PT-100" designates a pressure transmitter sequence numbered "100". Finally, the "A" suffix is used if there are two identical instruments within sequence number 100 — for example, when there is another pressure transmitter that has the number 100 in unit 1A, system AA. In that case, the second pressure transmitter will have the tag number "1A-AA-PT-100-B".

**Figure I-4: Complex Tag Number**

As shown below, the instrument tag number is comprised of five parts:

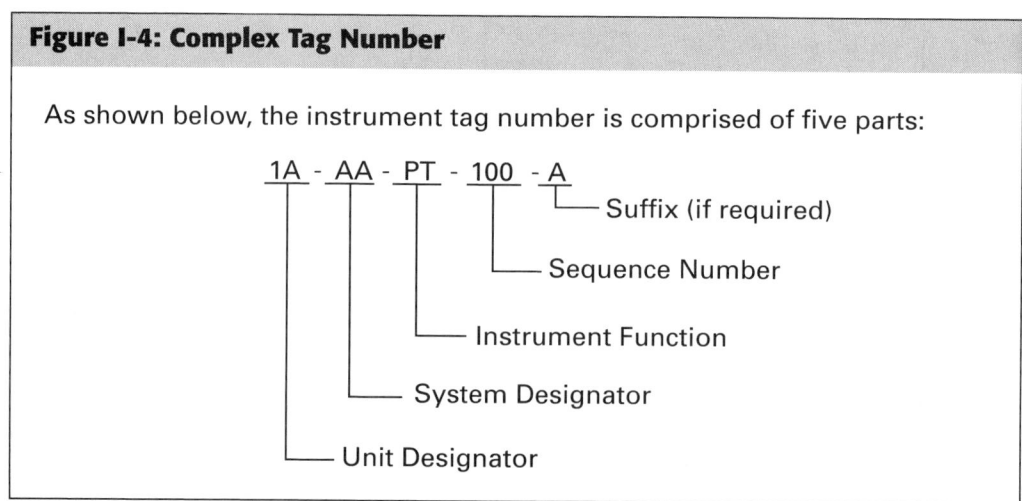

1A - AA - PT - 100 - A

└── Suffix (if required)

└── Sequence Number

└── Instrument Function

└── System Designator

└── Unit Designator

The instrumentation and controls design group personnel place tag numbers on the P&ID and enter them into the Instrument List or database for tracking. This is done for control purposes because, on a large project, there may be many, many P&IDs – perhaps one hundred or so – plus thousands of tag-marked devices. Since each device serves a specific function, all devices' status must be tracked until they are installed, their operation verified, and the plant has been accepted by the owner.

After tag numbers are entered on the Instrument List, the instrumentation and controls design group starts a Specification Form for each tag-marked item. Developing these Specification Forms is a major part of the instrument and control system design group's effort. Specification Forms must be completed to secure bids from suitable instrument suppliers, to purchase the items from the successful bidder, and to generate a permanent record of what was purchased.

As the design progresses, the need to define on-off or discrete control will become evident. For instance, on a pulp and paper mill project, it may be necessary to isolate a pump discharge to prevent pulp stock from dewatering in the pipe if the pump is shut down. An on-off valve is added to provide the isolation, but it is necessary to document why that device was added and what it is supposed to do. Since this on-off control affects other groups, it is important to define it as early and as accurately as possible. One way to do this is by using a Logic Diagram.

The instrumentation and controls group develops Installation Details based on the specific requirements of the devices it has specified, along with any owner-driven requirements. The installation requirements needed for good measurement are established by the instrument supplier, various industry groups and by the owners themselves. These requirements are then documented in the Installation Details. These details may be developed for the project, for the specific site, or possibly by the owner's corporate entity.

At the same time, the plant layout has also progressed so the instrumentation and controls design group can begin placing instruments on the Location Plans. These drawing are most often used to assist the construction contractor in locating the instruments, but they can also be useful for operations and maintenance because they show where instruments are installed in the completed plant.

Lastly, when all connection details are known and electric design has progressed to the point that wiring connection points are known, the instrumentation and controls design group can develop Loop Diagrams. These diagrams show all information needed to install and check out a loop. Because these diagrams may repeat information the piping, and electrical design teams included on their drawings, it is critically important that the instrumentation and controls group coordinates closely with other disciplines.

## Summary

In this introduction we have briefly described the documents that are included in instrumentation and control systems' set of deliverables and the sequence of their development. In the following chapters we will add more detail to describe the deliverables, how they can be used effectively, and how industry standards can assist in understanding.

Many illustrations in this book were originally developed for various ISA training courses, especially *ISA-FG-15 Developing and Applying Standard Instrumentaion & Control Documentation*, version 2.2, ISA standards, and other ISA publications. Origins of some illustrations are noted adjacent to the figure. The complete titles of the original documents are listed in Appendix B, Abbreviations. Some of the illustrations were revised for clarity and consistency.

## CHAPTER ONE

# The Process Flow Diagram, The PFD

The Process Flow Diagram (PFD) is a highly specialized document that you may actually have never seen. It is, nonetheless, critical to the organized, early development of any complex process. A PFD is the fundamental representation of a process that schematically depicts the conversion of raw materials to finished products without delving into details of how that conversion occurs. It defines the flow of material and utilities, it defines the basic relationships between major pieces of equipment, and it establishes the flow, pressure and temperature ratings of the process.

Project design teams use PFDs most effectively during the developmental stages of a project. During these stages feasibility studies and scope definition work are undertaken prior to commencing detailed design. PFDs are closely associated with material balances. They are used to decide if there are sufficient raw materials and utilities for a project to proceed. Within an operating company, a plant-wide design group and the site management may use PFDs to document the flow of process materials and utilities among the different units within a facility.

There is no generally accepted industry standard to aid in developing the PFD. Consequently, some PFDs show a minimum of detail while others may include significant detail. These two different design approaches are discussed below.

## Minimum Detail Approach

For a PFD to be effective, the entire process is shown in as little space as practical. Only the major process steps are depicted, and detail is minimized. The intent is to simply show a change has been made or a product has been produced, rather than how that change was made. It can be somewhat of a challenge to limit the detail shown on a PFD. For example, very little, or no, instrumentation and control (I&C) detail is shown on a PFD, since this equipment is not critical to the material balance. Nor are individual I&C components a significant cost component in the overall budget. Valves and transmitters are usually significantly less costly than an associated pressure vessel. Details will be shown later on the P&IDs and other project documents. P&IDs will be discussed in detail in the next chapter.

So, how do you decide what you show on your PFD? Well, if you are the I&C professional and you are using the minimum detail approach, not much of your work is used at this stage in the process development. One successful rule of thumb is to show detail on equipment only if that information has a significant impact on the material balance, or if that information is needed to define something special about that equipment. The term "special" here means a "significant cost impact to the project". If the information is needed to reach a critical project decision, it may be important enough to show on the PFD.

### Additional Detail Approach

Other plant design teams and plant owners believe a PFD should include more design details. These teams and owners involve the I&C engineers early in the project. The I&C engineers are involved in the development of the PFDs. The PFDs might then include design details such as major measurement points, control methods, control valves, and process analyzers. The PFDs are used as a guide, or perhaps even a first step, in the development of the P&IDs. Details will be shown (or duplicated) on the P&IDs and other project documents. P&IDs will be discussed in detail in Chapter 2.

A single PFD may contain enough information for several P&IDs. One rule of thumb is a PFD may contain enough information to develop up to 10 P&IDs!

The PFD's purpose is to define the design of the process. Figure 1-1 is an example of a simplified PFD. Completion of a PFD is frequently the starting point of the detailed engineering of a continuous process plant.

### Figure 1-1: Process Flow Diagram

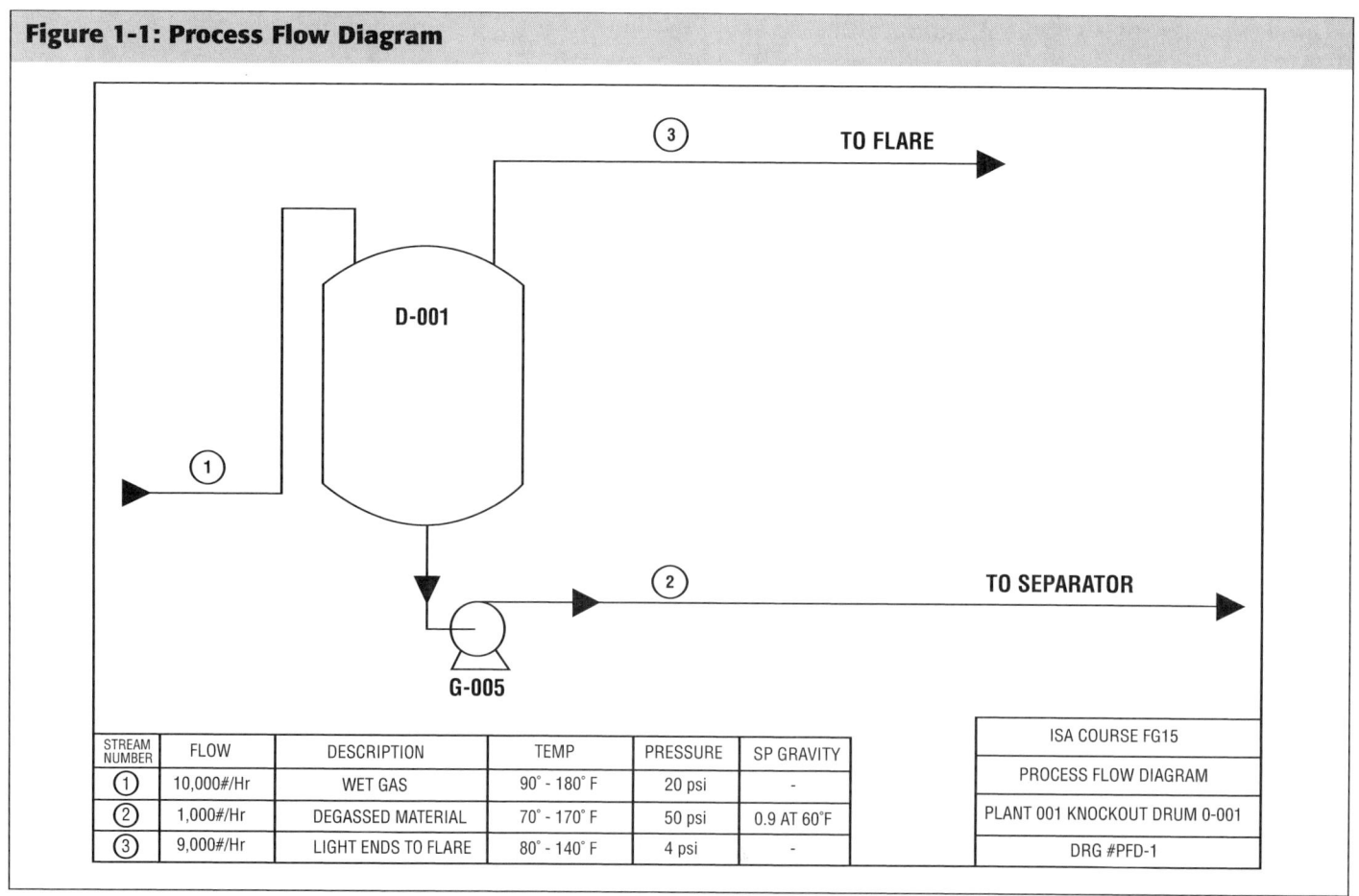

| STREAM NUMBER | FLOW | DESCRIPTION | TEMP | PRESSURE | SP GRAVITY |
|---|---|---|---|---|---|
| ① | 10,000#/Hr | WET GAS | 90° - 180° F | 20 psi | - |
| ② | 1,000#/Hr | DEGASSED MATERIAL | 70° - 170° F | 50 psi | 0.9 AT 60°F |
| ③ | 9,000#/Hr | LIGHT ENDS TO FLARE | 80° - 140° F | 4 psi | - |

| ISA COURSE FG15 |
|---|
| PROCESS FLOW DIAGRAM |
| PLANT 001 KNOCKOUT DRUM 0-001 |
| DRG #PFD-1 |

A PFD is most likely developed in several steps. The plant owner may develop a preliminary PFD, as a first step, to be used as a "thinking document" which sets down on paper a proposed process or a process change that is under consideration. The plant owner may elect to use other methods to document the work, such as a written description to define the process scope. See Figure 1-2, Process Description. In either form, this information is used to establish the initial design criteria for the plant.

The PFDs, or other conceptual information, is normally reviewed by the engineering contractor's process engineers and planning team before the release to detail design. The review is to ensure two criteria have been met:

1. There is enough information on the PFD to support development of the P&IDs by all the detail design disciplines. The decision that "enough" information is presented is probably best left to the design entity that will use the PFD.

2. Material balance information is present to support, with the experience of the project design and purchasing teams, identification and specification of "long lead" equipment. "Long lead" equipment is the equipment that requires a long time to procure, design, fabricate and ship. In other words, it is equipment that has to be purchased early in the project.

PFDs developed by a plant owner will likely be re-drawn by the engineering team. The new version will include the information needed by the design team.

The owner will put a lot of effort and invest a great deal of time, money and expertise in the project before any PFD is developed. The following is a simplified look at steps the owner will take.

A project may start with a gleam in someone's eye or a voice in the middle of the night. We could sell a lot more product if we had a new, more efficient plant. We could sell a new product like soap, or paint, or sodium bicarbonate, or tissue, or toluene di-isocyanate, or computer chips provided we could produce it in a cost-effective way. We could use a new plant, a new process, new materials, or different techniques. We could make our product better, or cheaper. We could reduce pollution, or have fewer by-products. We could make our product more profitable with higher quality. The gleam in the eye is then turned over to a team for further development.

### Figure 1-2: Process Description

• **Process Description Plant 001 Knockout Drum D-001**

  - The inlet gas, which consists of mixed petroleum liquids and vapors, originates in various sections of the plant and is piped to the knockout drum, D-001, where liquids and vapors are separated by expansion and a slow-down of velocity.

  - The mixed petroleum liquids are pumped to the separator and vapors are routed to the flare.

  - The incoming material is normally 10% condensate, but under some conditions, condensables may be reduced substantially.

  - The wet gas will vary in temperature from a low of 90°F to a high of 180°F.

The team will include company managers and specialists, such as consultants, engineers, real estate advisors, purchasing managers, marketing teams, sales experts, and other support personnel. The team develops, at the least, a general size and location for the plant, a marketing plan for the product, and a financial plan to establish and control costs. A preliminary process is defined with a PFD, and the source and costs of the raw materials are determined.

If all this information is favorable, company executives would likely decide to build a plant to make a specified number of units per year, using the best existing technology. The plan would possibly specify that the plant be located where raw materials, electricity, water and an intelligent labor force are available. The plan would have costs defined and escalation calculated for the project's duration. The cost plan would include the production yield forecasts as well as the planned cost of the raw materials, combined and massaged to provide a unit cost and margin for the units sold. The plan would ultimately project the return on investment (ROI) for the project, which hopefully will be above the company threshold for new projects. If there is little return on the investment, or if it is below the company threshold, the project is simply not going to be approved.

Planning continues after the decision is made to proceed with the project. Next, the executive team will secure the necessary land, and a set of scope definition documents will be completed. These will serve as the starting point for the detailed engineering. An initial or preliminary PFD, or other process description developed by the owner's engineers or consultants, is included in these scope documents. Many firms use independent engineering contractors for the detailed engineering. Other firms have in-house capabilities and staff and prefer to do the detailed engineering design themselves.

If an independent engineering contractor is to be used, the owner will use the scope documents to aid in securing the contractor's services through competitive bidding or by other selection processes.

A typical preliminary PFD, or process description, will show the product manufactured by the plant; raw materials necessary for that product; by-products produced by the process; waste materials that must be disposed of; process pressures, temperatures, and flows needed to produce the product; and major equipment needed. The important piping runs are shown, but piping is not sized on a PFD, and auxil-

## The Design Team

Whether a contractor develops the design, or it is done in-house, the work is done by an engineering design team, consisting of many specialty groups. A typical team will be led by a project engineer or engineering manager and it might consist of the following design groups:

| | |
|---|---|
| Civil | Process |
| Electrical | Project |
| Instrumentation and Control | Structural |
| Mechanical Equipment | Vessels |
| Plant Design/Piping | |

The design team is a part of the total organization necessary to manage the design and construction of a facility. One common term for the scope of the total organization is EPC: Engineering - Procurement - Construction. Some owners hire contractors for some or all of the three parts, while others handle all three themselves. The owner's project manager has overall control of the project. The project manager may also have additional staff to handle other functions, such as cost engineering, estimation and legal. Contractors may also use a project manager to control their portion of the project, if they have responsibilities other than engineering.

## Figure 1-3: PFD Equipment Symbols

**Subgroup:** Process
**Symbol Name:** Vessel
**Symbol Mnemonic:** VSSL

**Description**: A vessel or separator. Internal details may be shown to indicate type of vessel. Can also be used as a pressurized vessel in either a vertical or horizontal arrangement.

**Subgroup:** Process
**Symbol Name:** Distillation Tower
**Symbol Mnemonic:** DTWR

**Description:** A packed or trayed distillation tower used for separation. Packing or trays may be shown to indicate type of distillation tower.

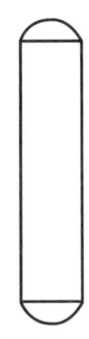

**Subgroup:** Storage
**Symbol Name:** Atmospheric Tank
**Symbol Mnemonic:** ATNK

**Description**: A tank for material stored under atmospheric pressure.

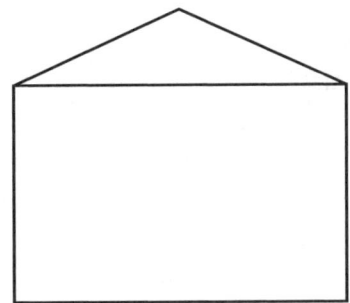

**Subgroup:** N/A
**Symbol Name:** Exchanger
**Symbol Mnemonic:** XCHG

**Description**: Heat transfer equipment. An alternative symbol is depicted.

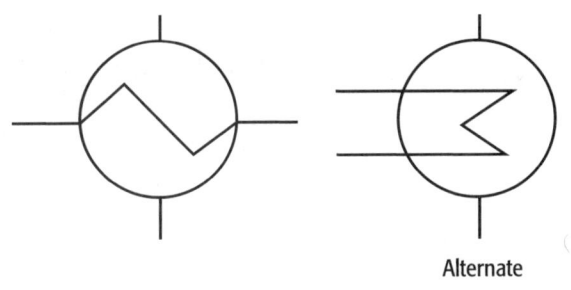

Alternate

*From ISA-5.5*

iary and utility piping are not shown. A written description of the process may also be included, perhaps to emphasize certain critical characteristics of the process.

The PFD will use symbols and letter designations to identify the equipment on the PFD. It is not necessary to add much detail to the equipment shown on a PFD. A simple line sketch will serve. For instance, a heat exchanger can be shown as a simple line representation of a main process flow and a heat transfer medium flow, without implying a particular type of exchanger. For a PFD, the only information needed is that a piece of equipment transfers heat at that point, rather than showing specifically the mechanism for transfer. For a few typical PFD symbols for equipment, see Figure 1-3.

Some projects might identify equipment by using the Symbol Mnemonics shown on Figure 1-3: VSSL for vessels, DTWR for distillation towers, ATNK for atmospheric tank, and XCHG for exchanger. Other projects might use a single letter for identification: such as, C for columns and tanks, D for drums and vessels, E for heat exchangers and coolers, and G for pumps. There are many variations of the letters and symbols used. It is very important to be consistent throughout a project, and almost as important to use symbols familiar to those who will use them.

The successful engineering contractor for the project will review and probably revise or replace the owner's PFD, or process definition, with a new PFD using the contractor's standards. It is likely to be more efficient for the contractor to redraw the PFDs to take advantage of their "standardized" symbol and drawing development features inherent in the contractor's computer-aided drafting (CAD) package.

Process flow data and conditions are provided on the PFD. These conditions are normally the "design" conditions, but — if it is important to the material balance or equipment sizing — normal or operating conditions, maximum conditions, and even minimum conditions may be provided. Since the PFD is tied closely to the material balance, mass flow units are normally used. Additionally, pressure and temperature conditions are provided as well.

There are two common ways to show the process information. One is to provide a set of numbers above, and possibly below, the line connecting equipment, using a standard format: flow/pressure/temperature. Delimiters are used between the conditions. Units are not provided normally to conserve space. The units are standardized and are provided in a legend sheet. The flow conditions are those upon which the project is based, the equipment is purchased, and the piping is sized later in the design process.

Another useful way to document process conditions is to use a keyed table. A numbered symbol — frequently a diamond with an internal number — is added above a line or piece of equipment on the drawing. A table is then provided along the top or bottom of the PFD, listing the process conditions for that numbered symbol. This approach has the advantage of simplifying the addition of additional process conditions, and makes it a bit easier to maintain data on the table.

As discussed earlier in this chapter, some engineering contractors or owners include more information on PFDs than the minimum described above. This should be agreed upon between the owner and the contractor. Arguably, when there is pressure to add more detail to the PFDs, it may well be time to redirect the design effort to P&IDs. Some projects may show basic or even more detailed instrumentation and controls information. However, very simple symbols are typically used to indicate these devices on a PFD.

## Batch Processing Plants Vary

Batch processing plants may contain equipment used in different ways, in different sequences - often for many different batches or products at one time, or at different times.

The PFD defines a continuous process very efficiently. Batch processing, however, may require additional definition. A batch process subjects a fixed quantity of material (a batch) to one or more process steps in one or more pieces of equipment. The process takes place in a set of equipment defined in ANSI/ISA-88.01-1995, Batch Control Part 1: Models and Terminology as a process cell.[1]

The process cell may be used to make a single product or many products. There are two further choices if the cell is making many products. The cell may use different raw materials with different process parameters and either use the same equipment or, alternatively, use different equipment. Many process cells have the capability to process more than one batch of the same, or different, products concurrently. A single PFD can define one process. In batch processing the PFD is often supplemented by a recipe, due to the complexity. Recipes contain five categories of information, as indicated in Figure 1-4, and are specific for the end product.

### Figure 1-4: Recipe Contents

| | |
|---|---|
| Header | Administrative information and a process summary |
| Equipment Requirements | Information about the specific equipment necessary to make a batch or a specific part of the batch |
| Procedure | Defines the strategy for carrying out a process |
| Formula | Describes recipe process inputs, process parameters, and process outputs |
| Other information | Product safety, regulatory, and other information that doesn't fit in the other categories |

Figure 1-4 is from the book *Applying S88, Batch Control from a User's Perspective*, written by Jim Parshall and Larry Lamb. The book contains a definition of a control recipe: "A control recipe is used to create a single specific batch…. Control Recipes unique to individual batches allow product tracing or genealogy to occur." [2]

Some engineering contractors or owners use the PFD as a first step in designing the instrumentation and control systems. Important process monitoring and control requirements are captured, as they become known. In this situation, the process design team will indicate on the PFD where various process variables are to be measured. For example, a circle with a single letter P inside signifies that the pressure at this point is important to the process and should be measured. Likewise, the use of F for flow, L for level, or T for temperature in a circle would indicate where these variables are measured. The exact instrumentation and control systems required would be developed later and shown on a P&ID.

Other contractors or owners might elect to show important, critical, or, most commonly, expensive instrumentation and control system components. For example, an in-line process chromatograph may appear on the PFD, either due to its importance to the overall process or because of its cost. Other project teams may elect to define process variable sensing points and show controllers and control valves symbolically. PFDs are intended to provide a canvas for the broad-brush artistry of the process engineers. The fine details wanted by the instrumentation and controls engineers should be left to the P&ID.

We have not shown any instrumentation symbols on our sample PFD, but we will discuss symbols and identification of the I&C systems in Chapter 2.

We have chosen a very simple continuous process for our discussion. We will develop the rest of design documents for our plant in the following chapters. The PFD for our simulated project is shown as Figure 1-1 and a word description of the process is shown as Figure 1-2.

The PFD in Figure 1-1 shows there is a flow in the process line, stream number (1), of 10,000 pounds/hour of wet gas with a temperature between 90°F and 180°F and a pressure of 20 psi. The variation in temperature is caused by process changes upstream of our PFD. Note that only a stream number, (1), (2) or (3) identifies the pipelines. Not included are line size, material of construction, or pressure rating (ANSI 150, ANSI 300, etc) for any of the piping shown on the PFD. Also note that there are no symbols or data shown for the pump driver. Only its equipment number, G-005, identifies the pump.

The wet gas goes into D-001, the Knockout Drum, where the liquid condenses out of the wet gas stream as the gas expands and cools. The liquid is pumped to a separator (on another PFD) where the water and process liquid are separated. Stream number (2) shows the pump G-005 has a discharge pressure of 50 psi. The pumped liquids have a specific gravity of 0.9 at 60°F. The pump has a capacity of 1,000 pounds/hour and the temperature of the degassed material varies between 70°F and 170°F.

The light ends or gases, 9,000 pounds/hour and shown as stream number (3), are piped to a flare, which is shown on another PFD. The pressure needed to move this quantity of gas to the flare is 4 psi. From this simple simulated PFD we have enough information to start development of the P&ID. To the project design team, the PFD becomes less important as the P&ID develops and the process temperatures, pressures, and flows are used to develop design criteria. However, if it is kept current as the project develops, it may be used to familiarize the contractor's and the owner's personnel with the process. It is usually far easier to understand the basics of a process from a PFD than from the P&IDs.

1. ANSI/ISA-88.01-1995, Batch Control, Part 1: Models and Terminology (Research Triangle Park, NC: ISA — The Instrumentation, Systems, and Automation Society, 1995) p. 22.

2. Jim Parshall and Larry Lamb, *Applying S88 Batch Control From a User's Perspective* (Research Triangle Park, NC: ISA - The Instrumentation, Systems, and Automation Society, 2000) p. 48.

## CHAPTER TWO

# P&IDs and Symbols

## Overview

The acronym "P&ID" is widely understood within the process industries as the name for the principal document used to define a process – the equipment, piping and all monitoring and control components. *The Automation, Systems and Instrumentation Dictionary*, 4th edition's definition for a Piping and Instrumentation Drawing (P&ID) tells what they do. P&IDs "show the interconnection of process equipment and the instrumentation used to control the process".[1]

Sets of symbols are used to depict mechanical equipment, piping, piping components, valves, equipment drivers and instrumentation and controls. These symbols are assembled on the drawing in a manner that clearly defines the process. The instrumentation and control (I&C) symbols used in P&IDs are generally based on ISA-5.1-1984-(R1992), Instrumentation Symbols and Identification.[2]

This book will aid in solving the long existing and continuing problem of confusing information on P&IDs. The fact that there is confusion can be understood because there really is no universal standard that specifies what information should be included on a P&ID or even, for that matter, the meaning of the letters P&ID. You may know exactly what "P" means, or what "D" means or what a P&ID contains, but the person in the facility down the road probably doesn't agree. For instance, the "P" in P&ID may stand for Piping or Process. The "I" refers to Instrument or Instrumentation. The "D" is for Drawing or Diagram. P&IDs may even be called "Flow Diagrams", which are not to be confused with Process Flow Diagrams discussed in the previous chapter. P&IDs are sometimes called "Flow Sheets", a term often preceded by the department that initiated or developed them, like "Engineering", or "Controls", or other descriptors. In this book, for simplicity, we will refer to the document by the acronym, P&ID.

There is no universal, national or international, multi-discipline standard that covers the development and content of P&IDs. However, much of the information and use of a P&ID is covered by ISA-5.1, which is an excellent, flexible document that defines, primarily, instrument symbolism.

This book uses ISA-5.1 as the definitive reference. We are aware the document is under review and revision at this writing, in early 2004. Some changes will probably be included when the revision is issued, but we are sure the intent and focus of the standard will be maintained.

Another professional organization, Process Industry Practices (PIP), has developed and published many recommended practices. Among these is one on P&IDs. There is additional information about PIP in Chapter 10.

The P&IDs in your facility have probably been produced and revised over many years by many different developers. Many different individuals have documented revisions to the content – and even the symbolism – of your P&IDs to reflect process improvements and additions, as well as changing control technology. Unless you have been incredibly and unbelievably fortunate maintaining your site standards, some of your P&IDs will use symbolism and format that differ from the original and even from each other. As you well know, inconsistent symbolism and format of your P&IDs can be annoying, confusing, and more importantly, it makes information they contain subject to misunderstandings.

Although the P&ID is the overall document used to define the process, the first document developed in the evolution of a process design is often the PFD, the Process Flow Diagram, discussed in Chapter 1. Once a PFD is released for detail design, the project scope has been established and P&ID development may commence.

P&IDs develop in steps. The key members of the design team – perhaps plant design, piping, process, and project specialists, all lay out a conceptual pass at showing vessels, equipment and major piping. The instrumentation and controls are typically added next, since they often require significant additional space on the P&ID. Or, in the words of one project manager, "you guys sure do have lots of bubbles". Then, the contributions of the specialists in electrical, mechanical equipment, vessels and other disciplines are added. These specialists fill in the information blocks containing equipment numbers, titles and definitive text reserved for critical information regarding the equipment: size, rating, throughput, and utility usage (horsepower). The developmental process is an iterative one. Information is added in steps until the document is complete with all necessary details.

P&IDs are controlled documents formally issued at various stages. Control means changes to the drawings are identified and clearly documented in some manner and there is verification checking or some other quality assurance procedure in effect. The care needed to control the content of P&IDs can be understood in light of the fact that P&IDs carry the definitive information from which many design entities draw their work. From the P&ID comes the Instrument List and the specification, acquisition and installation of all instrumentation and controls. From the P&ID comes the motor list with horsepower. From the P&ID come the piping line list, sizes, service and purpose. The drawings even document critical information regarding tanks, vessels and other equipment – all of which are used to lay out equipment and start the specification and purchasing efforts. In some states, P&IDs carry professional engineers' stamps.

P&IDs are distributed to members of the project team and interested client personnel after quality control checking and under rigorous revision control. This formal issue process will occur several times in the course of a project. The drawings are so important that key milestones are often built into the project schedule based on the different issues of P&IDs. Some typical formal P&ID drawing issues may include:

A – Issue for scope definition

B – Issue for Client Approval

C – Issue for bid, bidding of major equipment

D – Issue for detailed design

0 – Issue for construction (or 1, or 2, or 3, etc.)

Before we start looking at a P&ID we shall define a few terms, with particular focus on instrumentation and controls.

Figure 2-1 contains a few simple definitions. An instrument is a device for measuring, indicating, or controlling a process. This includes both simple and complex devices. Pressure gauges or dial thermometers are typical simple ones. Complex devices may include process analyzers – perhaps a gas chromatograph, which defines types and quantities of gases in a process stream.

> **Figure 2-1: Instrument & Process Control Defined**
>
> - **Instrument**
>   - A device for measuring, indicating, or controlling
> - **Process control**
>   - All first-level control – process or discrete – consists of three parts:
>     - Sensing
>     - Comparing
>     - Correcting

The term "Process Control" can be understood from any dictionary definition of the two words. In its simplest form, a process is a series of steps and control is to regulate. So process control is the regulation of a series of steps.

All types of process control include three functions: sensing, comparing and correcting.

## Sensing

First, we have to know where we are by sensing the relevant characteristic of our environment – otherwise known as the process. One definition of process sensing is to ascertain or measure a process variable and to convert that value into some understandable form (see Figure 2-2).

---

**Figure 2-2: Sensing & Comparing Defined**

- **Sensing**

  To ascertain or measure a process variable and convert that value into some understandable form

- **Comparing**

  To compare the value of the process variable (PV) with the desired value set point (SP) and to develop a signal to bring the two together. The signal depends on:

  - How far apart the PV & SP are
  - How long they have been apart
  - How fast they are moving toward or away from each other

The flow of fluid in a pipe or air in a duct, the level of liquid in a tank, the pressure of gas in a vessel, the temperature of the fluid inside a distillation tower are all process variables. Normally, in process control, these variables are measured continuously. A transmitter measures the process in some way and transmits the information to a central location where the comparison takes place. The central location is usually a control room where plant operators monitor the process, or, for purists, the rack room where the process control computer is located that performs the comparison.

## Comparing

Figure 2-2 contains a formal definition of the comparing function. The value of the process variable is compared with the desired value (the set point), and action is taken to develop a signal to bring the two together. The control is automatic and continuous. Comparison takes place in a pneumatic or electronic controller or via a shared display shared control system, such as a distributed control system (DCS), a programmable logic controller (PLC), a computer chip embedded in a field instrument, or even a desktop computer. These devices may look at three characteristics of the process:

P-Proportional or gain – how far away the process variable is from the set point

I-Integral or reset – how long the process variable has been away from the set point

D-Derivative or rate – how fast the process variable is changing

It is just coincidental that the three components of a process control algorithm yield the same acronym (PID) as the primary design drawing that details the process under control.

## Correcting

The control device then develops a signal to bring the process variable and the set point together. This signal is transmitted to a field device that changes the value of the process variable. This device is most often a control valve or a variable speed pump drive. See Figure 2-3.

---

**Figure 2-3: Correcting Defined**

- **Correcting**

  —To bring the process variable closer to the set point. This is accomplished by the final control element – most often this is a *control valve*

- **Control valves, usually, but not always:**

  – Are pneumatically actuated, often by a 3-15 psi signal

  – Can be moved directly by a pneumatic controller

  – Are actuated by a transducer if the controller signal is electronic or digital

---

## The Control Loop

In automatic control, the three devices – the transmitter that senses, the controller that compares, and the control valve that corrects – are interconnected to form a control loop. The interconnection may be pneumatic, electronic, digital, or a combination of all three. The pneumatic component is typically a 3-15 psig (pounds per square inch gauge) instrument air signal. If the interconnection is electronic, a 4-20 mA (milliamperes) signal is usually used, although other signal levels are also used. The signal level is a function of the control system selected. As yet, there is no agreement in industry on a digital transmission standard, and entire books are written on the relative merits of the various protocols.

---

**Figure 2-4: Loop Defined**

A combination of interconnected instruments that measures and/or controls a process variable

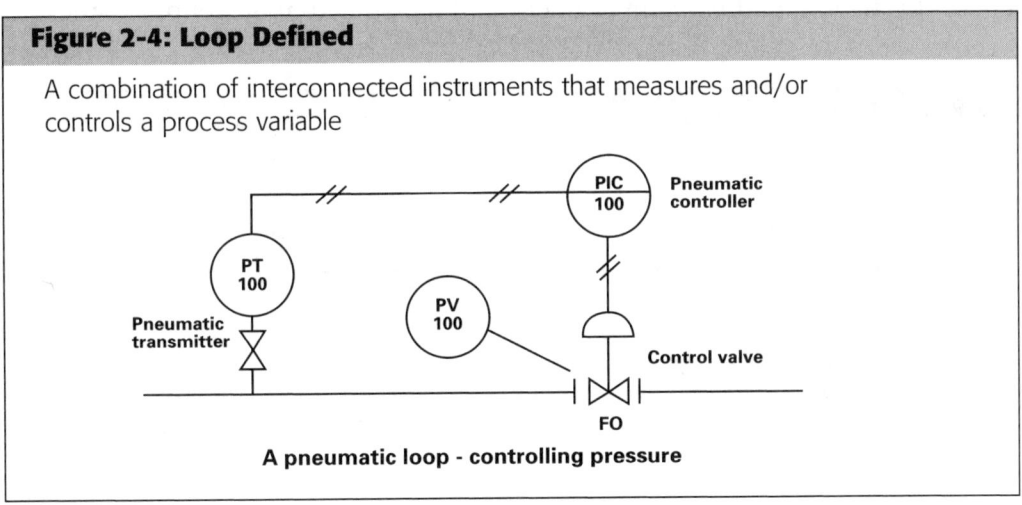

**A pneumatic loop - controlling pressure**

---

Figure 2-4 shows a pneumatic loop controlling the pressure in a pipeline. The loop number is 100, so all the devices in the loop will have the number 100. The double crosshatched lines indicate information is transmitted pneumatically from the transmitter PT-100 to the indicating controller PIC-100, and from PIC-100 to the control valve PV-100 with a signal varying from a low of 3 psig to a high of 15 psig. The control valve moves according to the value of the 3-15 psig signal. It has a FO identifier, meaning that if the primary power source to the valve is lost, in this case pneumatic pressure, the valve will Fail Open.

## What's Missing?

Is the drawing of the simple pressure loop complete? There probably is no right answer to that – other than, "What do you think?". We are not really ducking the question. But remember, the people responsible for the P&ID will have to live with the drawing for many years. The "stakeholders" in the project need to decide how much detail is provided on a P&ID. The intended uses of the P&ID as a design document, a construction document and to define the system for operations all will, in some way, influence the detail shown. A list of a few things that might be shown include:

*Air sets* – Sometimes a symbol is added to pneumatic devices that indicates where instrument air is connected and an air set is needed. The air set is made up of any combination of a pneumatic regulator, a filter and a pressure gauge.

*Set points* – Some firms add the set points for regulators and switches, although we believe these are better shown on a Loop Diagram.

*Root valve* – The instrument root valve between the process and the transmitter may have a size and specification called out.

*Control valve size* – Sometimes the size of the valve is inferred by the size of the piping or by the size of piping reducers; sometimes the size is provided as a superscript outside the instrument bubble.

*Valve positioner* – In our opinion, the use of a valve positioner can be defined in the construction and purchasing specifications and Installation Details. We see no need to show positioners on the P&ID.

*Controller location* – The panel, bench board, control room or other location can be added as an identifier outside, but near to, the controller bubble. These will usually appear as an acronym or as a few letters that are further identified on the P&ID legend sheet.

## Control Valves

Control valves may fail in various positions – open, closed, locked, or indeterminate. The position of a failed valve can have a significant impact on associated equipment, and, therefore, it is of interest to operations personnel. Valve fail action is often discussed and agreed upon during the P&ID review meetings, so it is natural and efficient to document the agreed-upon action on the

### Figure 2-5: Actuator Action and Power Failure

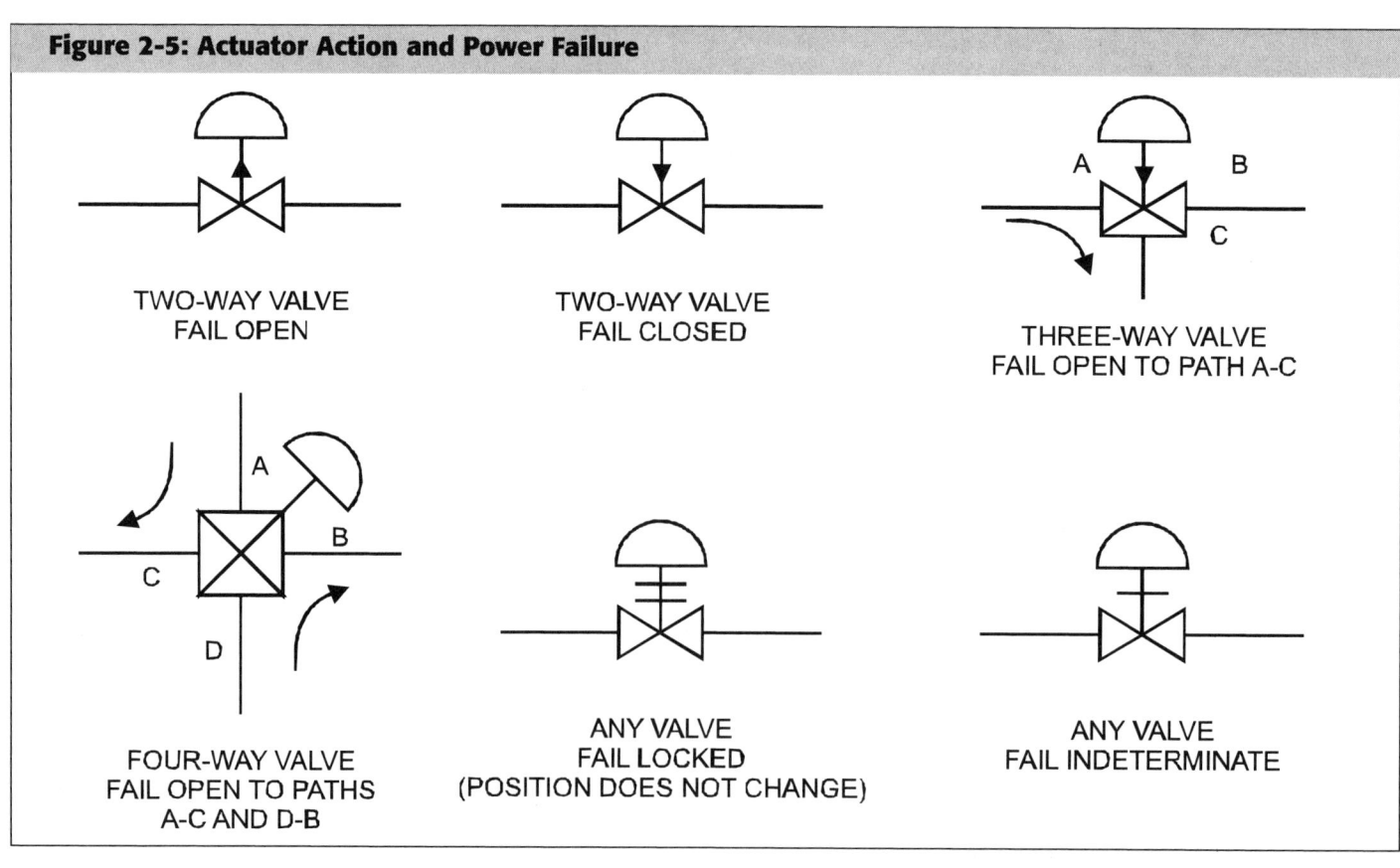

TWO-WAY VALVE
FAIL OPEN

TWO-WAY VALVE
FAIL CLOSED

THREE-WAY VALVE
FAIL OPEN TO PATH A-C

FOUR-WAY VALVE
FAIL OPEN TO PATHS
A-C AND D-B

ANY VALVE
FAIL LOCKED
(POSITION DOES NOT CHANGE)

ANY VALVE
FAIL INDETERMINATE

P&ID. For valve fail action, the term "Power" means the medium that moves the valve actuator and therefore the valve trim. The most common "Power" medium is instrument air. Power does not refer to the signal, unless the signal is the medium that moves the actuator.

The fail positions may be identified on the P&ID using letters below the valve symbol: FO for Fail Open; FC for Fail Closed; FL for Fail Last or Locked; and FI for Fail Indeterminate. Figure 2-5, Actuator Action and Power Failure, shows other methods of indicating the fail position of control valves. Looking at the figures, an arrow up signifies the valve fails open. An arrow down is fail close. A crossing line is fail indeterminate. Two crossing lines indicate fail locked or last position.

It is important to remember that fail position refers to the loss of the primary power at the valve, the motive force. Pulling the electronic signal off the valve transducer or electro-pneumatic positioner may induce a different reaction than the failure indication shown. A springless piston actuated valve will fail indeterminate upon loss of air. However, if there is a positioner, it will be driven in one direction or the other upon loss of the electronic signal.

### Natural Gas Can Substitute for Air

Pneumatic systems are not always pressurized by instrument air. Offshore hydrocarbon production platforms have a ready supply of compressed gas available, albeit natural hydrocarbon gas. For smaller platforms without electric power, a gas filter dryer serves quite well in preparing the pneumatic medium to control the platforms. Obviously, smoking at work is frowned upon.

The control panels are a complex mass of pneumatic tubing, containing specialized components like first-out pneumatic indicators called "winkies". Natural gas doesn't have a noticeable smell. The familiar rotten egg smell of natural gas is actually due to a stenching agent – an odorant added later as a safety feature for consumers. It's a very effective solution to a specific challenge.

### Figure 2-6: An Electronic Loop

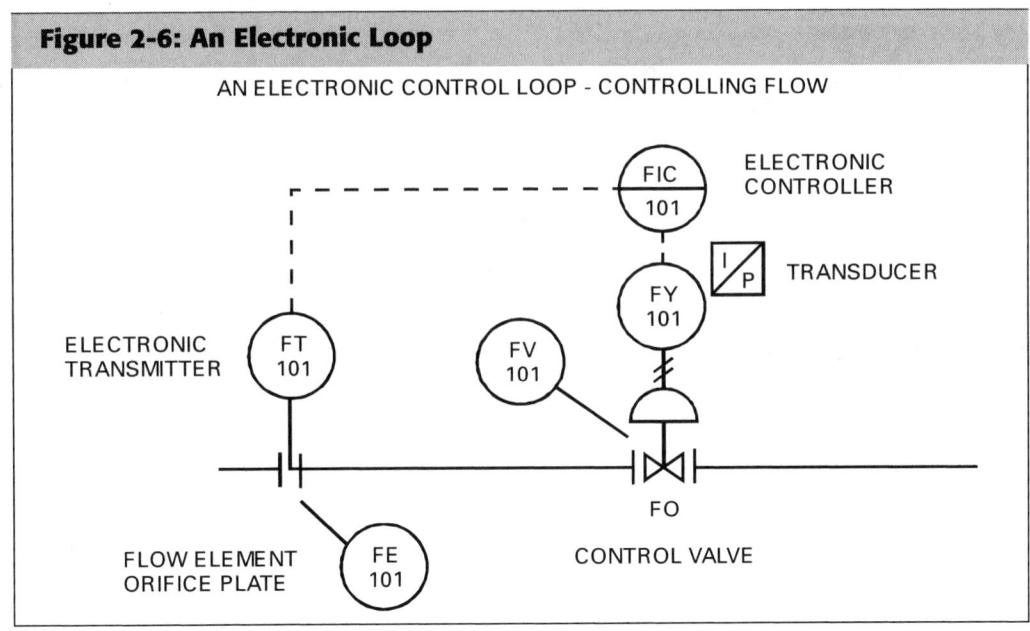

AN ELECTRONIC CONTROL LOOP - CONTROLLING FLOW

Figure 2-6 shows an electronic loop controlling flow in a pipeline. The loop number is 101. The dotted line indicates that information is transmitted electronically from the flow transmitter, FT-101, to the indicating controller, FIC-101, and from the controller to the current to pneumatic converter (I/P), FY 101. FT-101 senses the differential pressure proportional to the flow rate in the line caused by FE-101, a flow element or orifice plate. FT-101 transmits a 4-20 mA dc (direct current) signal corresponding to the varying differential pressure. FIC-101, an electronic flow indicating controller, transmits a 4-20 mA dc signal to the converter or transducer, FY-101, that converts the 4-20 mA dc signal into

## To Show or Not to Show?

One of the challenges you will face is the depiction of third party systems on your P&IDs. If you have an island of equipment furnished by a third party, how much of that equipment should show on your drawing? If the third party system suppliers have their own P&IDs, do you copy them into your drawing set, or possibly just include their P&ID with your set? As usual, there really is no right answer; each facility is managed differently, each project has a different scope and each stakeholder in the P&IDs has different requirements.

It is not inexpensive to redraw a P&ID within your drawing set, nor is it a particularly good idea to have two drawings that show the same thing – yours, and the system supplier's. The drawings will probably only agree on the day they are checked and issued for use. As soon as someone makes a change, you start to "chase revisions".

One successful and cost effective approach has been to show the interface points between the vendor's system and your control system – just show the components seen on the operator station. Then, on your drawing, refer to the vendors P&ID and operating manual for further details.

a pneumatic signal. This signal changes the position of the valve actuator, which in turn changes the position of the inner works of the control valve, changing the flow through the control valve.

Simple instruments permit direct reading of a process variable in the field. These devices include pressure gauges, thermometers, level gauges and rotameters. Other loops are slightly more complex, transmitting a signal to the remote control system to indicate or record the value of a process variable in the control room, but without a controlled output. Both these classes of instruments are shown on a P&ID.

Members of the instrumentation and control design group add all the loop and local instruments to the P&ID, one at a time, until the complete instrumentation and control system is defined on the drawing. The proper location of local instruments should not be neglected, as they can be the first line of contact for those running and maintaining the facility. Your facility can only be improved when the operators and maintenance personnel assist with the endeavor.

### ISA-5.1

ISA-5.1 is the standard most often used in process industries as the basis for depicting instrumentation and control systems on P&IDs and other documents. It is broad in scope and flexible in usage. The following is a quote from ISA-5.1, paragraph 4.4.1, entitled Symbols.

"The examples in this standard illustrate the symbols that are intended to depict instrumentation on diagrams and drawings. Methods of symbolization and identification are demonstrated. The examples show identification that is typical for the pictured instrument or functional interrelationships. The symbols indicate the various instruments or functions have been applied in typical ways in the illustrations. This usage does not imply, however, that the applications as designations of the instrument or functions are restricted in any way. No inference should be drawn that the choice of any of the schemes for illustration constitutes a recommendation for the illustrated methods of measurement or control. Where alternative symbols are shown without a statement of preference, the relative sequence of symbols does not imply a preference."[3]

The basic process control tagging standard for most industrial facilities is based on ISA-5.1. You will find, however, that additional information or interesting interpretations have been added to further define local requirements, to meet specific system requirements, or even to maintain site tradition. It is critically important that the standards used at your facility are completely defined and rigidly followed.

Without careful control of the symbols and usage, your documentation will rapidly devolve into a mess that is difficult to understand and use. More importantly, when the drawings are confusing to read or difficult to work with, people simply stop using them. Drawings and documentation must be continuously updated to agree with improvements and additions. When there is any problem with using the drawings, if they are confusing, ambiguous, difficult to read, or inaccessible, they will not be maintained. Drawings that are not maintained with vigilance quickly become useless, or worse, inaccurate.

## Device Definition

As can be seen from Figures 2-4 and 2-6, a combination of identification letters, numbers, and symbols is used to define the devices in a loop. The identification letters are specified in ISA-5.1 and reproduced as Figure 2-7.

Figure 2-7 consists of twenty-six rows and five columns. The first column lists, alphabetically, twenty-six process variables, or as ISA-5.1 states, the "measured or initiating variable".[4] The first letter of any tag name, therefore, will indicate the process variable being measured. The most common process variables in a process plant include:

F – Flow

L – Level

P – Pressure

T – Temperature

There are several letters — C, D, G, M, N, O, which can be specified by the user. Of course, the user must clearly document the specified meanings on the site P&ID legend sheet, and those meanings should be maintained, without ambiguity or change, for the entire facility or, ideally, the entire company. Many sites will use ISA-5.1 as the starting point. The legend sheet table can then be modified to incorporate assigned letter designations, or even specifically define acceptable or standard letter combinations for the facility.

Using X for the first letter is a special case. From ISA-5.1, "The unclassified letter X is intended to cover unlisted meanings that will be used only once or used to a limited extent. If used, the letter may have any number of meanings."[5] The function of the letter is defined both on the legend sheet as well as implied with a few descriptive letters adjacent to the bubble. When properly applied, the letter X does not appear frequently – only once, or to a limited extent. Instead, the user-defined letters should be used for devices that appear regularly, even if infrequently. Thus, in many modern industrial facilities, X may not be needed, since most devices appear with some regularity. For those of you that have an entire facility filled with XT transmitters or XY transducers,

## Figure 2-7: Identification Letters

| | FIRST-LETTER (4) | | SUCCEEDING-LETTERS (3) | | |
|---|---|---|---|---|---|
| | MEASURED OR INITIATING VARIABLE | MODIFIER | READOUT OR PASSIVE FUNCTION | OUTPUT FUNCTION | MODIFIER |
| A | Analysis (5,19) | | Alarm | | |
| B | Burner, Combustion | | User's Choice (1) | User's Choice (1) | User's Choice (1) |
| C | User's Choice (1) | | | Control (13) | |
| D | User's Choice (1) | Differential (4) | | | |
| E | Voltage | | Sensor (Primary Element) | | |
| F | Flow Rate | Ratio (Fraction) (4) | | | |
| G | User's Choice (1) | | Glass, Viewing Device (9) | | |
| H | Hand | | | | High (7, 15, 16) |
| I | Current (Electrical) | | Indicate (10) | | |
| J | Power | Scan (7) | | | |
| K | Time, Time Schedule | Time Rate of Change (4, 21) | | Control Station (22) | |
| L | Level | | Light (11) | | Low (7, 15, 16) |
| M | User's Choice (1) | Momentary (4) | | | Middle, Intermediate (7,15) |
| N | User's Choice (1) | | User's Choice (1) | User's Choice (1) | User's Choice (1) |
| O | User's Choice (1) | | Orifice, Restriction | | |
| P | Pressure, Vacuum | | Point (Test) Connection | | |
| Q | Quantity | Integrate, Totalize (4) | | | |
| R | Radiation | | Record (17) | | |
| S | Speed, Frequency | Safety (8) | | Switch (13) | |
| T | Temperature | | | Transmit (18) | |
| U | Multivariable (6) | | Multifunction (12) | Multifunction (12) | Multifunction (12) |
| V | Vibration, Mechanical Analysis (19) | | | Valve, Damper, Louver (13) | |
| W | Weight, Force | | Well | | |
| X | Unclassified (2) | X Axis | Unclassified (2) | Unclassified (2) | Unclassified (2) |
| Y | Event, State or Presence (20) | Y Axis | | Relay, Compute, Convert (13, 14, 18) | |
| Z | Position, Dimension | Z Axis | | Driver, Actuator, Unclassified Final Control Element | |

From ISA-5.1

don't worry, this provision of ISA-5.1 is frequently ignored. You are not alone. Worry only if you are inconsistent!

The second column, marked "Modifier", adds additional information about the first letter, the process variable. For example, if an instrument is used to measure the difference between two pressures, perhaps the upstream and downstream pressure of a filter press, a P for pressure is used as the first letter and a D for differential as a second letter modifier. See Figure 2-8 and 2-9. When instantaneous flow is being measured and a totalizer is added to provide total flow over time, the device identification is FQ. The first letter of the tag name is F for flow and the second letter is Q from the second column, signifying integrate or totalize.

The next three columns further define the device. The first of these delineates a readout or passive function. For example, Figure 2-8 shows that the filter press differential pressure is measured and indicated, as shown by a third letter I, for indicator. The absence of a dividing line in the middle of the circle (or "bubble") shows the differential pressure is displayed locally.

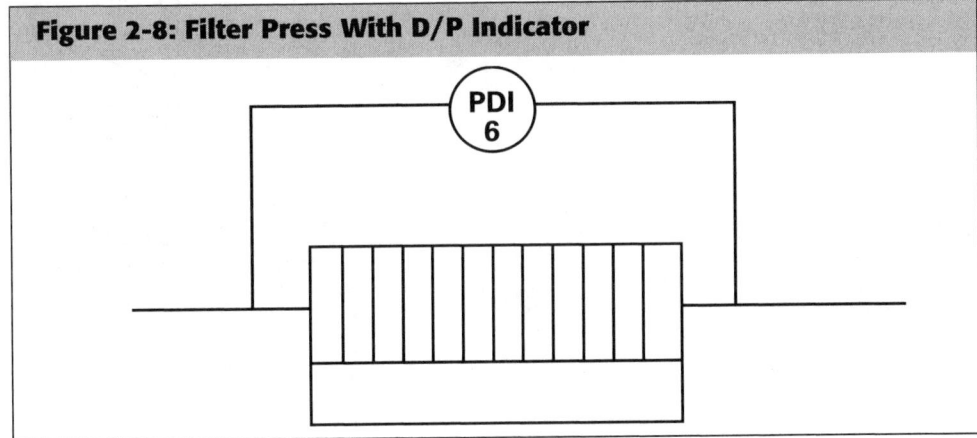

**Figure 2-8: Filter Press With D/P Indicator**

Therefore, PDI shows locally the pressure drop across the filter. Figure 2-9 shows that the pressure differential value is transmitted to a central location. The second column of succeeding letters shows that we would use a T for transmitter, so the device would be a PDT.

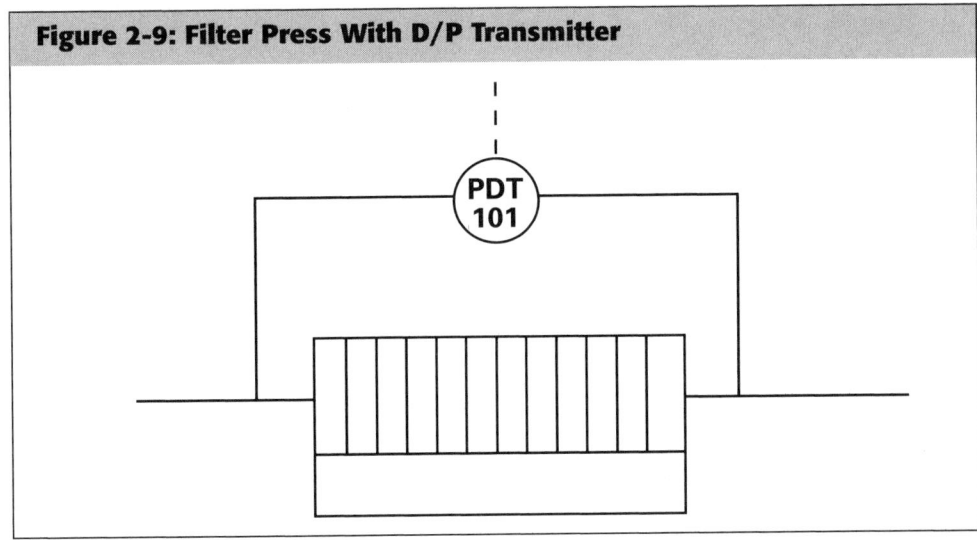

**Figure 2-9: Filter Press With D/P Transmitter**

## Figure 2-10: Typical Letter Combinations

| First-Letters | Initiating or Measured Variable | Controllers Recording | Controllers Indicating | Controllers Blind | Self-Actuated Control Valves | Readout Devices Recording | Readout Devices Indicating | Switches and Alarm Devices* High** | Switches and Alarm Devices* Low | Switches and Alarm Devices* Comb | Transmitters Recording | Transmitters Indicating | Transmitters Blind | Solenoids, Relays, Computing Devices | Primary Element | Test Point | Well or Probe | Viewing Device, Glass | Safety Device | Final Element |
|---|---|---|---|---|---|---|---|---|---|---|---|---|---|---|---|---|---|---|---|---|
| A | Analysis | ARC | AIC | AC | | AR | AI | ASH | ASL | ASHL | ART | AIT | AT | AY | AE | AP | AW | | | AV |
| B | Burner/Combustion | BRC | BIC | BC | | BR | BI | BSH | BSL | BSHL | BRT | BIT | BT | BY | BE | | BW | BG | | BZ |
| C | User's Choice | | | | | | | | | | | | | | | | | | | |
| D | User's Choice | | | | | | | | | | | | | | | | | | | |
| E | Voltage | ERC | EIC | EC | | ER | EI | ESH | ESL | ESHL | ERT | EIT | ET | EY | EE | | | | | EZ |
| F | Flow Rate | FRC | FIC | FC | FCV, FICV | FR | FI | FSH | FSL | FSHL | FRT | FIT | FT | FY | FE | FP | | FG | | FV |
| FQ | Flow Quantity | FQRC | FQIC | | | FQR | FQI | FQSH | FQSL | | | FQIT | FQT | FQY | FQE | | | | | FQV |
| FF | Flow Ratio | FFRC | FFIC | FFC | | FFR | FFI | FFSH | FFSL | | | | | | FE | | | | | FFV |
| G | User's Choice | | | | | | | | | | | | | | | | | | | |
| H | Hand | | HIC | HC | | | | | | HS | | | | | | | | | | HV |
| I | Current | IRC | IIC | IC | | IR | II | ISH | ISL | ISHL | IRT | IIT | IT | IY | IE | | | | | IZ |
| J | Power | JRC | JIC | JC | | JR | JI | JSH | JSL | JSHL | JRT | JIT | JT | JY | JE | | | | | JV |
| K | Time | KRC | KIC | KC | KCV | KR | KI | KSH | KSL | KSHL | KRT | KIT | KT | KY | KE | | | | | KV |
| L | Level | LRC | LIC | LC | LCV | LR | LI | LSH | LSL | LSHL | LRT | LIT | LT | LY | LE | | LW | LG | | LV |
| M | User's Choice | | | | | | | | | | | | | | | | | | | |
| N | User's Choice | | | | | | | | | | | | | | | | | | | |
| O | User's Choice | | | | | | | | | | | | | | | | | | | |
| P | Pressure/Vacuum | PRC | PIC | PC | PCV | PR | PI | PSH | PSL | PSHL | PRT | PIT | PT | PY | PE | PP | | | PSV, PSE | PV |
| PD | Pressure, Differential | PDRC | PDIC | PDC | PDCV | PDR | PDI | PDSH | PDSL | PDSHL | PDRT | PDIT | PDT | PDY | PE | PP | | | | PDV |
| Q | Quantity | QRC | QIC | QC | | QR | QI | QSH | QSL | QSHL | QRT | QIT | QT | QY | QE | | | | | QZ |
| R | Radiation | RRC | RIC | RC | | RR | RI | RSH | RSL | RSHL | RRT | RIT | RT | RY | RE | | RW | | | RZ |
| S | Speed/Frequency | SRC | SIC | SC | SCV | SR | SI | SSH | SSL | SSHL | SRT | SIT | ST | SY | SE | | | | | SV |
| T | Temperature | TRC | TIC | TC | TCV | TR | TI | TSH | TSL | TSHL | TRT | TIT | TT | TY | TE | TP | TW | | TSE | TV |
| TD | Temperature, Differential | TDRC | TDIC | TDC | TDCV | TDR | TDI | TDSH | TDSL | TSHL | TDRT | TDIT | TDT | TDY | TE | TP | TW | | | TDV |
| U | Multivariable | | | | | UR | UI | | | | | | | UY | | | | | | UV |
| V | Vibration/Machinery Analysis | | | | | VR | VI | VSH | VSL | VSHL | VRT | VIT | VT | VY | VE | | | | | VZ |
| W | Weight/Force | WRC | WIC | WC | WCV | WR | WI | WSH | WSL | WSHL | WRT | WIT | WT | WY | WE | | | | | WZ |
| WD | Weight/Force, Differential | WDRC | WDIC | WDC | WDCV | WDR | WDI | WDSH | WDSL | WDSHL | WDRT | WDIT | WDT | WDY | WE | | | | | WDZ |
| X | Unclassified | | | | | | | | | | | | | | | | | | | |
| Y | Event/State/Presence | YRC | YIC | YC | | YR | YI | YSH | YSL | | YRT | YIT | YT | YY | YE | | | | | YZ |
| Z | Position/Dimension | ZRC | ZIC | ZC | ZCV | ZR | ZI | ZSH | ZSL | ZSHL | ZRT | ZIT | ZT | ZY | ZE | | | | | ZV |
| ZD | Gauging/Deviation | ZDRC | ZDIC | ZDC | ZDCV | ZDR | ZDI | ZDSH | ZDSL | ZDSHL | ZDRT | ZDIT | ZDT | ZDY | ZDE | | | | | ZDV |

**Note:** This table is not all-inclusive.
*A, alarm, the annunciating device, may be used in the same fashion as S, switch. the actuating device.
**The letters H and L may be omitted in the undefined case.

**Other Possible Combinations:**

| | |
|---|---|
| FO | (Restriction Orifice) |
| FRK, HIK | (Control Stations) |
| FX | (Accessories) |
| TJR | (Scanning Recorder) |
| LLH | (Pilot Light) |
| PFR | (Ratio) |
| KQI | (Running Time Indicator) |
| QQI | (Indicating Counter) |
| WKIC | (Rate-of-Weight-Loss Controller) |
| HMS | (Hand Momentary Switch) |

From ISA-5.1

By starting with a process variable at the left of Figure 2-9 and adding the letters defined in the succeeding columns, the complete function of the control system device is defined.

Figure 2-10, Typical Letter Combinations, a reprint of a page in ISA-5.1[6], shows many possible letter combinations and describes the device represented by the letters. Reading across Figure 2-10, starting with an F, the initiating or measured variable for flow rate, the succeeding letters describe the devices and functions as follows:

| Letter Combination | Description |
| --- | --- |
| FRC | Flow Recorder Controller. A recorder for the value of instantaneous flow, integral with a flow controller. |
| FIC | Flow Indicating Controller. An instantaneous flow indicator combined with a flow controller. |
| FC | Blind Flow Controller. A flow controller without any indication or recording of instantaneous flow. |
| FCV | A self-actuated control valve controlling flow. |
| FICV | An FCV with an integral instantaneous flow indicator. |
| FR | Flow Recorder. |
| FI | Flow Indicator. |
| FSH | Flow Switch High. A switch which changes state on high flow. |
| FSL | Flow Switch Low. A switch which changes state on low flow. |
| FSHL | Flow Switch High-Low. A switch which changes state on high or low flow, and does not change in between the high and low flows. |
| FRT | Flow Recording Transmitter. For transmitting and recording in the same device. |
| FIT | Flow Indicating Transmitter. Transmitter with an integral indication of instantaneous flow. |
| FT | Blind Flow Transmitter. Transmitter with no indication of instantaneous flow. |
| FY | Solenoid, Relay, Computing Device. For example, current (I) to pneumatic (P) converters are correctly identified (in accordance with ISA-5.1) as FY in a flow loop, with a further definition of I/P shown outside the symbol, often in a square box. |

| Letter Combination | Description |
| --- | --- |
| FE | Primary Element. An orifice plate. |
| FP | Test Point. A point provided in the piping where a test measurement is made; the instrument is not normally connected to the point permanently. The point is normally valved or otherwise isolated. |
| FG | Flow Glass. G for glass or viewing device. A sight flow indicator. An uncalibrated view of the flow is provided. |
| FV | Flow Valve. Control valve in a flow loop. |

## Interesting interpretations – and an opinion:

An electro-pneumatic transducer, commonly called an I/P, probably has more combinations of "ISA standard" tags that any other control system component. We say "ISA standard" somewhat facetiously, since clearly all approaches cannot be correct, yet you can be sure that someone along the way assured someone else that their particular approach was in accordance with ISA-5.1. Creative tagging of I/Ps include, even within a single facility, no tag at all, I/P, IP, FY, XY, NY and so forth. "No tag" can easily occur when the I/P arrives on site pre-mounted to a control valve by the valve vendor, and the control valve is the only tag within your system.

The correct tagging of an I/P is to use the first letter of the loop in which the I/P appears, "the process variable", followed with a "Y" as the output function "convert". Thus a flow loop I/P would be an FY. To be crystal clear, the I/P would be written in a function block or a box adjacent to the bubble. The reason for the creative tagging of I/P may be that, with the widespread use of electronic instrument databases, some may see an advantage in developing a unique identifier for an I/P, so a database sort can list all the I/Ps on a project under one identifier independent of the loop it serves.

There will be a lot of I/Ps on a project. The ability to list all occurrences of a component is handy when specifying and purchasing a component. Also, from a practical standpoint, this author was once asked, "since they are called I/Ps, why not tag them as "I/Ps"? It is hard to argue with that logic! The I/P tag works since there is not another common device that would call for the use of I/P; there isn't a data clash.

Detailed explanations to justify the I/P tag typically start with: "I is the process variable for current and P is pneumatic pressure, so it works." Well, that may be true. It certainly works, but it isn't technically true from an ISA-5.1 view, so some practitioners may be appalled. The "process variable" letter is intended for the entire loop, not for that one device in the loop, so it should technically be F, P, T, L, etc., the variable that the loop is measuring or controlling. "P" is not actually listed as an output function. P is "pressure" only as a process variable, the first letter in the tag string.

## Instrument Numbering

In addition to the letters, the instrumentation and control design group assigns a sequence number to each function. All the devices within that function carry the same sequential number – in other words, the loop number. A single loop number is used to identify the devices that accomplish a single specific action – usually an input and an output for PID control, an input for indication of a process variable, or a manual output. This number, combined with the letter designation, positively and uniquely identifies each device within that set.

These numbers may follow the suggestions in ISA-5.1. However, there are many other numbering systems used in industry. ISA-5.1 suggests that loop numbering may be parallel or serial. By parallel, ISA-5.1 means starting a new number sequence for each first letter. Therefore, there may be an FRC-101, a PIC-101, and a TI-101. By serial, ISA-5.1 means using a single numerical sequence for all devices. Therefore, there may be an FRC-101, a LR-102, a PIC-103, and a TI-104.

A block of numbers is sometimes used to designate certain types of devices. For example, all safety valves might use the 900 series: PSV-900, PSV-901, PSV-902, etc.

Instrument numbers may also be structured to identify the loop location or service. For example, see Figure 2-11, Instrument Numbering. The first digit of the number may indicate the plant number; hence, FT-102 is an instrument in plant 1. Another method of identifying the instrument location is with a prefix, for example: 2 (area), or 03 (unit), or 004 (plant 4) which identifies the service area of the loop: 2-FT-102 is loop 102 in area 2, or 03-FT-102 is

> **Figure 2-11: Instrument Numbering**
>
> - **Use Basic Number if project is small and there are no area, unit, or plant numbers:**
>   - Basic Number  FT-2 or FT-02 or FT-002
> - **If project has a few areas, units, or plants (9 or less), use the first digit of the plant number as the tag number:**
>   - FT-102   (1 = area, unit, or plant number)
> - **If project is divided into area, units, or plants:**
>   - 1-FT002
>   - 01-FT002
>   - 001-FT002

loop 102 in unit 03, or 004-FT-102 is loop 102 in plant 4. These numbers can also be combined to show area-unit-plant in one number: 234-FT-102 is a flow transmitter in loop 102, which serves area 2, unit 3 and plant 4. To be completely confusing, remember that the loop number defines the items in the loop, so the loop may serve the area listed above, but a particular device may be physically located in another area.

A variation of this system is to tie the P&ID numbers to a particular area, and then to sequentially number the instruments on that P&ID sheet. For example, P&ID 25 carries up to 100 loops, or instrument loop numbers 2500 to 2599. The elegance of this system is that you can find the correct P&ID for an instrument based upon the tag number alone, since the tag number includes the P&ID number. Frequently the area number is nested in the P&ID number anyway, so you will also know the area served by the loop just by looking at the loop number.

The numbering system chosen for your P&IDs and loops should be tested and verified to ensure they work as expected with the various electronic applications used within your facility. Loop number 1 and loop number 001 will have markedly different treatment by some databases and by the maintenance planning and inventory control software.

Many different numbering systems are used. Some incorporate a major equipment number into the instrument identification. Still another variation deviates from the "unique number" requirement by use of the "loop" number as a coding system to group similar commodity type devices. The number that appears in the loop number place on the instrument circle is a component identifier that is tied to a master device specification. This approach can be useful for calling out devices that don't interface to the control system, such as local indicators like pressure gauges – in other words, commodity type devices.

For example, in your facility, PI-100 is listed in the component's specification as a 4 1/2" diameter pressure gauge with a range of 0-150 psig and a stainless steel Bourdon tube. As long as all your PI-100s are the same, this system works. There will be many component numbers when this system is used on temperature gauges, since there are so many variations of stem length, dial size and ranges. When you have a different material of construction or other change, a different number has to be used. Of course, a more complete pressure gauge specification can be used when actually purchasing the gauge. This approach is not common, but with care, it can be useful.

The letters and numbers that identify loop components have to appear on a drawing somewhere, so the next step is to put the chosen function identifier and loop number, the tag number, on the P&ID. ISA-5.1 provides the information needed to further define the location of an instrument and controls device through the use of specific symbols. The symbols are shown in Figure 2-12.

**Figure 2-12: General Instrument or Function Symbols**

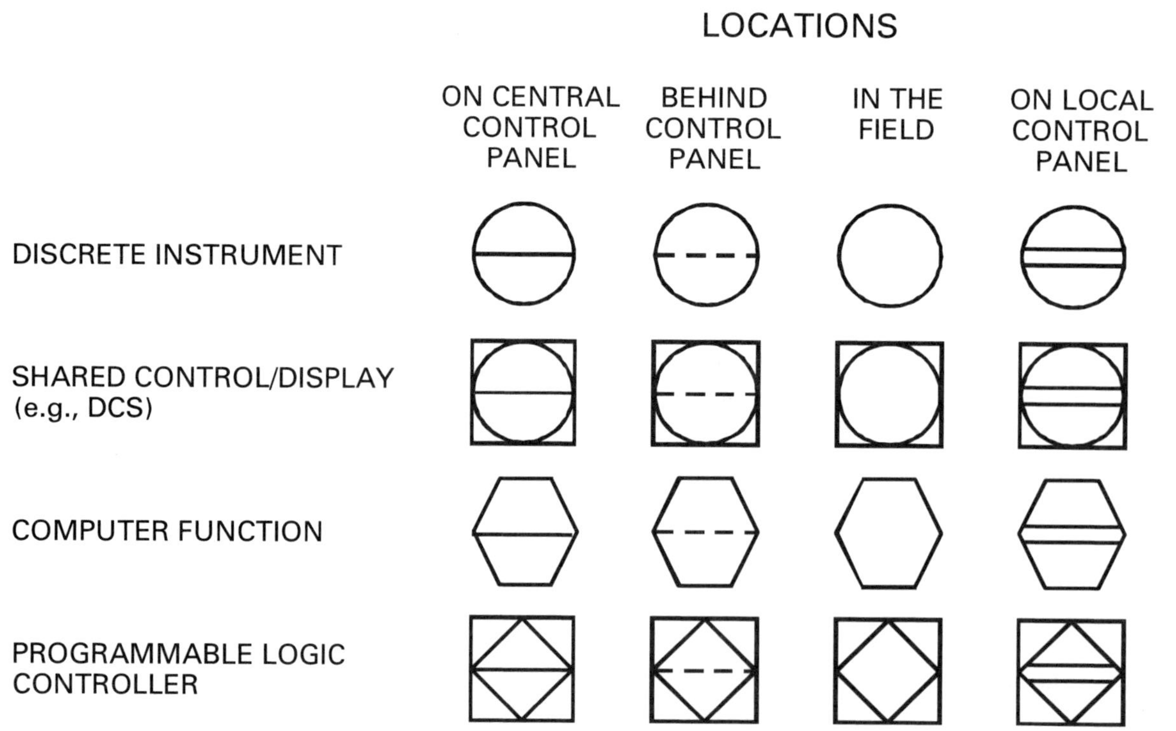

The circles, squares, hexagons and diamonds all have meaning. A circle means the device is field mounted (located in the process area of the plant). If a line is added through the center of the circle, the device is located in the primary location normally accessible to the operator (the central control room). If a second line is added, parallel to the first, the device is located in an auxiliary location, normally accessible to the operator (a local panel or on the starter cassette in the motor control center). A dotted line through the center of the circle shows the device is normally inaccessible to the operator (behind the panel). If an external square is added to the circle, the symbols represent devices or functions that are part of a shared display shared control system (a distributed control system, a DCS). If we substitute a hexagon for the circle or the squared circle, the symbols represent a computer function. A diamond within a square is used to define functions within a programmable logic controller, a PLC.

Line symbols are used to define the ways information is transferred between the field devices and the central control location.

## Line Symbols

Figure 2-13, Instrument Line Symbols, is copied from ISA-5.1.[7] The symbols describe how signals are transmitted between devices. First, the lines used are to be lighter than the associated process piping. The process sensing line, the pipe or tubing that connects a pressure transmitter directly to the process, is the lightest acceptable "pipe" line. A line with a double parallel crosshatch defines pneumatic transmission – usually, but not always, instrument air (some gas pipelines use natural gas, some plants use nitrogen).

For binary or on-off pneumatic signals, an optional symbol calls for the addition of a single opposing strike on the crosshatch of the pneumatic signal line. This is not commonly used, if for no other reason than the additional information regarding the nature of the signal is probably not critically important to the P&ID. Two symbols are shown for electronic transmission – the dotted line and the triple crosshatch. In the United States, the dotted line is predominant. Further definition of the electronic signal as binary or discrete is available by adding an opposing strike thru across the electric line. However, this electric binary symbol is not commonly used. Unguided electromagnetic transmission – including heat, radio waves, nuclear radiation and light – is shown by a series of sine waves. If the sine waves are superimposed on a line, the waves are guided. Internal system links such as software or data link are shown as a dash and a circle. This symbol is commonly used for a digital signal.

The advent of digital communication to field devices introduces the option to use either a dashed electric signal line or a data link line type to connect field transmitters and valve controllers. One approach may be to keep the dashed line symbol for field device wiring and the line-circle-line symbol to define

---

**Figure 2-13: Instrument Line Symbols**

ALL LINES TO BE FINE IN RELATION TO PROCESS PIPING LINES.

(1)  INSTRUMENT SUPPLY  *
     OR CONNECTION TO PROCESS

(2)  UNDEFINED SIGNAL

(3)  PNEUMATIC SIGNAL **

(4)  ELECTRIC SIGNAL                                                OR

(5)  HYDRAULIC SIGNAL

(6)  CAPILLARY TUBE

(7)  ELECTROMAGNETIC OR SONIC SIGNAL ***
     (GUIDED)

(8)  ELECTROMAGNETIC OR SONIC SIGNAL ***
     (NOT GUIDED)

(9)  INTERNAL SYSTEM LINK
     (SOFTWARE OR DATA LINK)

(10) MECHANICAL LINK

OPTIONAL BINARY ( ON-OFF ) SYMBOLS

(11) PNEUMATIC BINARY SIGNAL

(12) ELECTRIC BINARY SIGNAL                                         OR

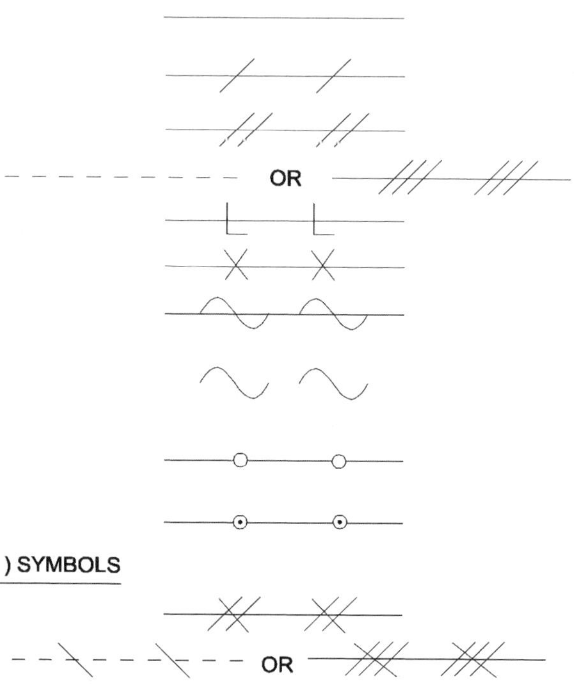

NOTE:  'OR' means user's choice.  Consistency is recommended.

   *  The following abbreviations are suggested to denote the types of power
      supply.  These designations may also be applied to purge fluid supplies.

      AS - Air Supply                          HS  - Hydraulic Supply
         IA  - Instrument Air          }Options NS  - Nitrogen Supply
         PA - Plant Air                         SS  - Steam Supply
      ES - Electric Supply                      WS - Water Supply
      GS - Gas Supply

      The supply level may be added to the instrument supply line, e.g., AS-100,
      a 100-psig air supply; ES-24DC, a 24-volt direct current power supply.

  **  The pneumatic signal symbol applies to a signal using any gas as the
      signal medium.  If a gas other than air is used, the gas may be
      identified by a note on the signal symbol or otherwise.

 ***  Electromagnetic phenomena include heat, radio waves, nuclear radiation,
      and light.

*From ISA-5.1*

---

function relationships within the control computer or for main data links
between control computers. A line with a single crosshatch is an undefined
signal, perhaps to be used in the early development stage of a P&ID.

## Pneumatic Transmission

### Figure 2-14: Pneumatic Transmission

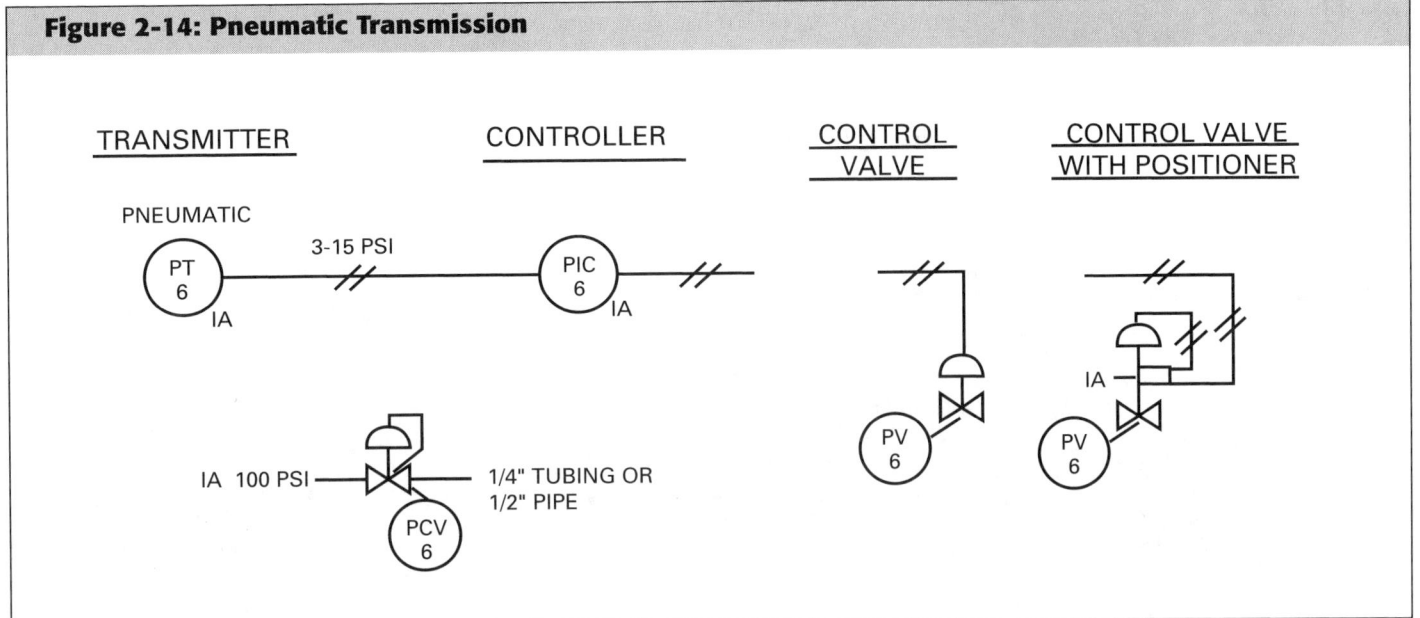

A complete pneumatic transmission system is shown in Figure 2-14. For the purposes of this example, pneumatic signal pressures are 3-15 psig. In practice, signal pressures can also be 6-30 psig, albeit less commonly. PT-6, a field mounted pressure transmitter, develops and transmits a 3-15 psig signal proportional to pressure of the process. The signal is transmitted to a field-mounted indicating controller, PIC-6. The controller develops and transmits the 3-15 psig corrective signal to the control valve PV-6.

If the valve operator (actuator or top works) can move the control valve through its entire range with the 3-15 psig signal, regardless of the process pressure, the pneumatic line is connected directly to the valve operator. If the 3-15 psig signal is not sufficient to operate the valve for all of its design conditions and range, a positioner is added to the valve operator. The function of the positioner is to compare the incoming signal with the actual valve position and develop the output air pressure necessary to position the valve in accordance with the incoming signal. The output pressure from the positioner to the valve is at a higher pressure, normally 30 psig up to and including full instrument air line pressure of 100 psig or higher.

Figure 2-14 shows that we need a source of instrument air (IA) at the transmitter, another at the controller, and another at the valve positioner. The IA is usually distributed in the field by a complete instrument air piping system, often at a nominal line pressure of 100 psig. Pressure regulators, shown in the figure as PCV-6, are located at the individual users to reduce the instrument air pressure to that required by the field device. Pressure regulators that serve pneumatic devices do not always, or – depending upon your industry – do not commonly carry loop identifiers. They may appear on the drawings as an untagged symbol, like a darkened triangle or some variation of an A with a line through it to the pneumatic device served.

### Electronic Transmission

Figure 2-15 shows a typical electronic transmission system.

**Figure 2-15: Electronic Transmission**

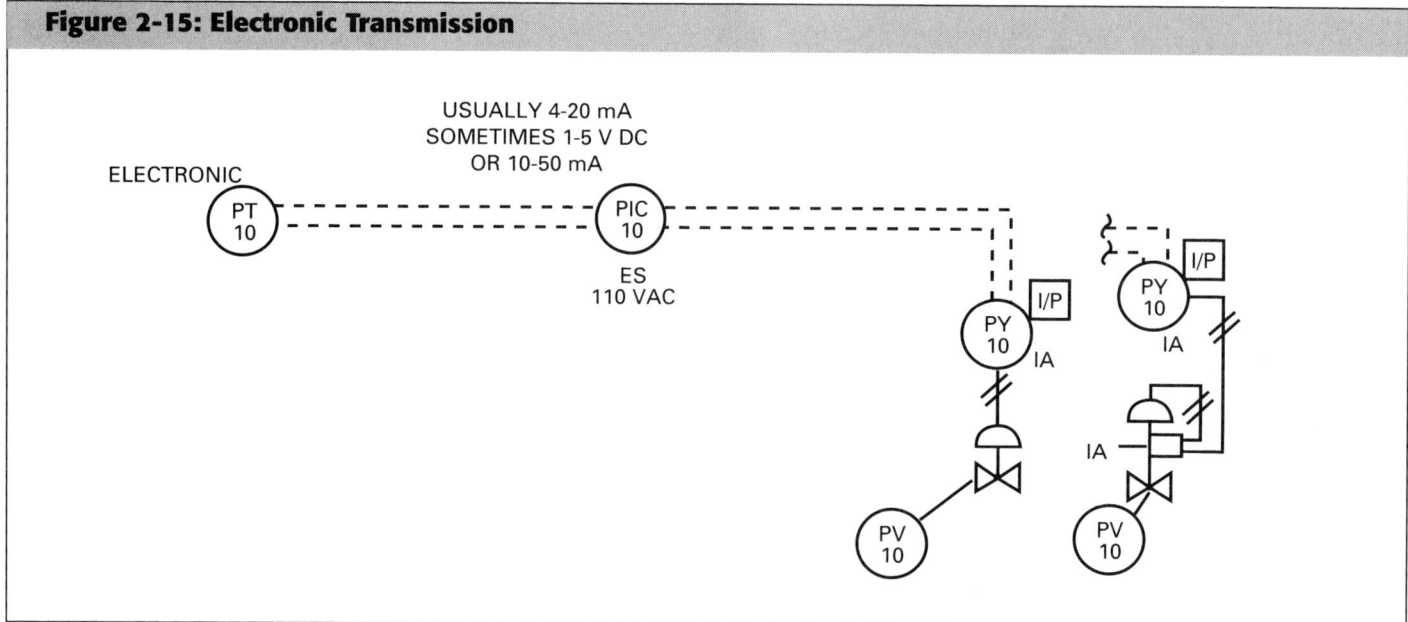

Many electronic transmission systems for instrumentation and controls are called "a two wire system". This means the field transmitter has only two wires connected to it. The signal transmitted usually has a 4 mA to 20 mA, nominally 24 volt dc range, although some installations may use a 10-50 mA dc or a 1-5 volt dc signal.

Most control valves are pneumatically operated, so even in the modern electronic control system, the electronic signal will be converted to pneumatic to actually change the position of the valve. The device that does this is a converter or transducer, typically an I/P or an electro-pneumatic positioner. An I/P is shown on Figure 2-15 as a PY, a traditional tagging convention. P is for pressure and Y is for solenoid, relay or computing device.

To clarify further, a function block, a small (1/4") square surrounding the letters I/P, is added to the right of the converter instrument circle. A pneumatic or electro-pneumatic positioner is frequently not tagged separately from the valve, probably because it is usually installed and shipped as part of the control valve. However, for your use, there is a symbol and a tag for positioners (ZC) included in ISA-5.1.[8] Symbolically, a simple box on the stem of a control valve can be used to indicate the presence of a positioner. An electro-pneumatic positioner is indicated when the electronic signal terminates on the box instead of a pneumatic signal.

There are many other symbols included in ISA-5.1 for specific instruments. We will not try to show them all. The following figure shows a few types of valves.

## Valves

The general valve symbol, the "bow tie", may be used to indicate the body of a control valve or a hand-operated valve. Some projects use this symbol as a generic control valve symbol rather than trying to define the control valve type by using the butterfly, globe or rotary symbols, shown in Figure 2-16.

**Figure 2-16: Valves**

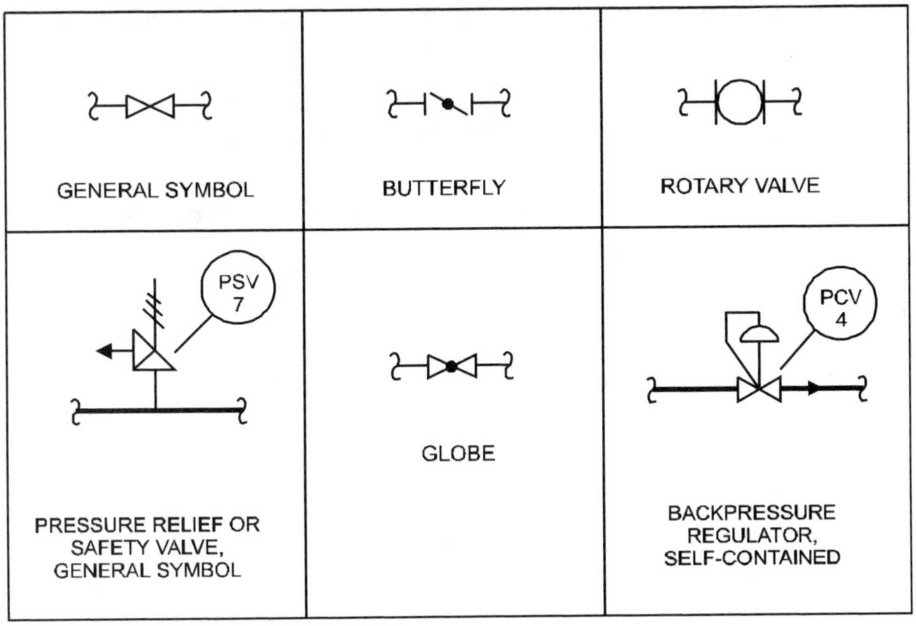

*From ISA-5.1*

It is important at this stage to balance the importance of the information with the expense of maintaining that information. In deciding to reflect the actual control valve type through the use of the specific valve symbol, you should ask if that information is important to the function of the drawing, since it serves your team and those that will make use of the drawing in the future.

For a P&ID, is it germane to understanding the process? Is it necessary to know that a control valve is a butterfly style when reviewing the P&ID? Do the preponderance of P&ID users care what kind of control valve is used, or is it enough to know there is a control valve there? The expense of the information is the cost to maintain the correct symbol. We are using the control valve symbol as one example of questions that should be asked when deciding what goes on a P&ID and what does not.

P&IDs are developed (rather than maintained), for the most part during a design project. The actual type of control valve may not be known until the valve is purchased, which is long after the P&IDs have been issued for detail design. You may be pretty sure a valve will be a butterfly valve, but you really won't know until the valve is purchased. If you are showing the actual valve type on the P&ID, someone will have to review each valve symbol after control valves are purchased to ensure the correct valve type symbol was chosen. There is a cost for that review, correction, and, even more so, to re-issue the drawings. On a large project, the cost of copying and distributing the drawings can be astronomical. Once the P&ID is issued and the project is complete, the details regarding that particular device are available elsewhere – on Loop Diagrams, Data Sheets, the Instrument Index, etc.

Additional valve symbols are shown in Figure 2-16. The symbol for safety or relief valves consists of an angle valve combined with a spring. Pressure regulators are control valves with actuators, but without an external control signal – designated in the example as a PCV, a self activated valve that regulates pressure. The pressure sensing line is shown upstream, if the PCV controls back pressure, and downstream if it controls downstream pressure.

One of the most common methods of measuring flow and transmitting that measurement is with an orifice plate and a differential pressure (d/p) cell.

**Figure 2-17: Typical Transmitters – Flow**
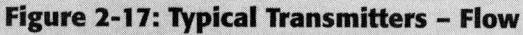

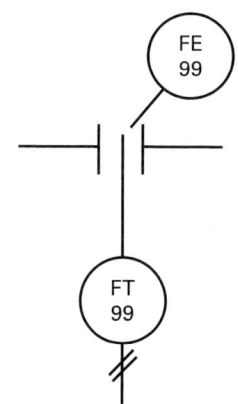

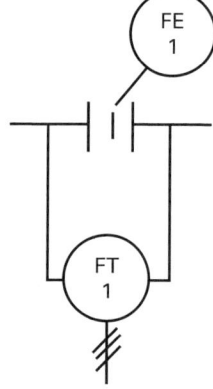

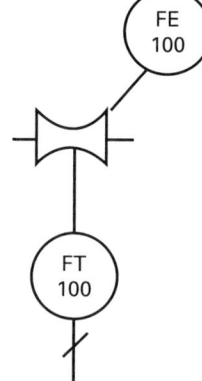

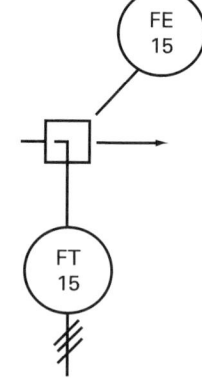

Orifice plate and orifice flanges with flange taps, differential pressure transmitter, pneumatic transmission

Orifice plate and flanges, taps are made in pipe, differential pressure transmitter, electronic transmission

Venturi tube; taps are in tube, differential pressure transmitter with indicator, undefined transmission

Pitot tube, connections are in tube, differential pressure transmitter, electronic transmission

Figure 2-17 shows several variations of primary flow elements that produce a pressure differential relative to flow: an orifice plate and flanges with flange taps, an orifice plate and flanges and pipe taps, a venturi tube, and a pitot tube.

## Figure 2-18:  Flow Devices

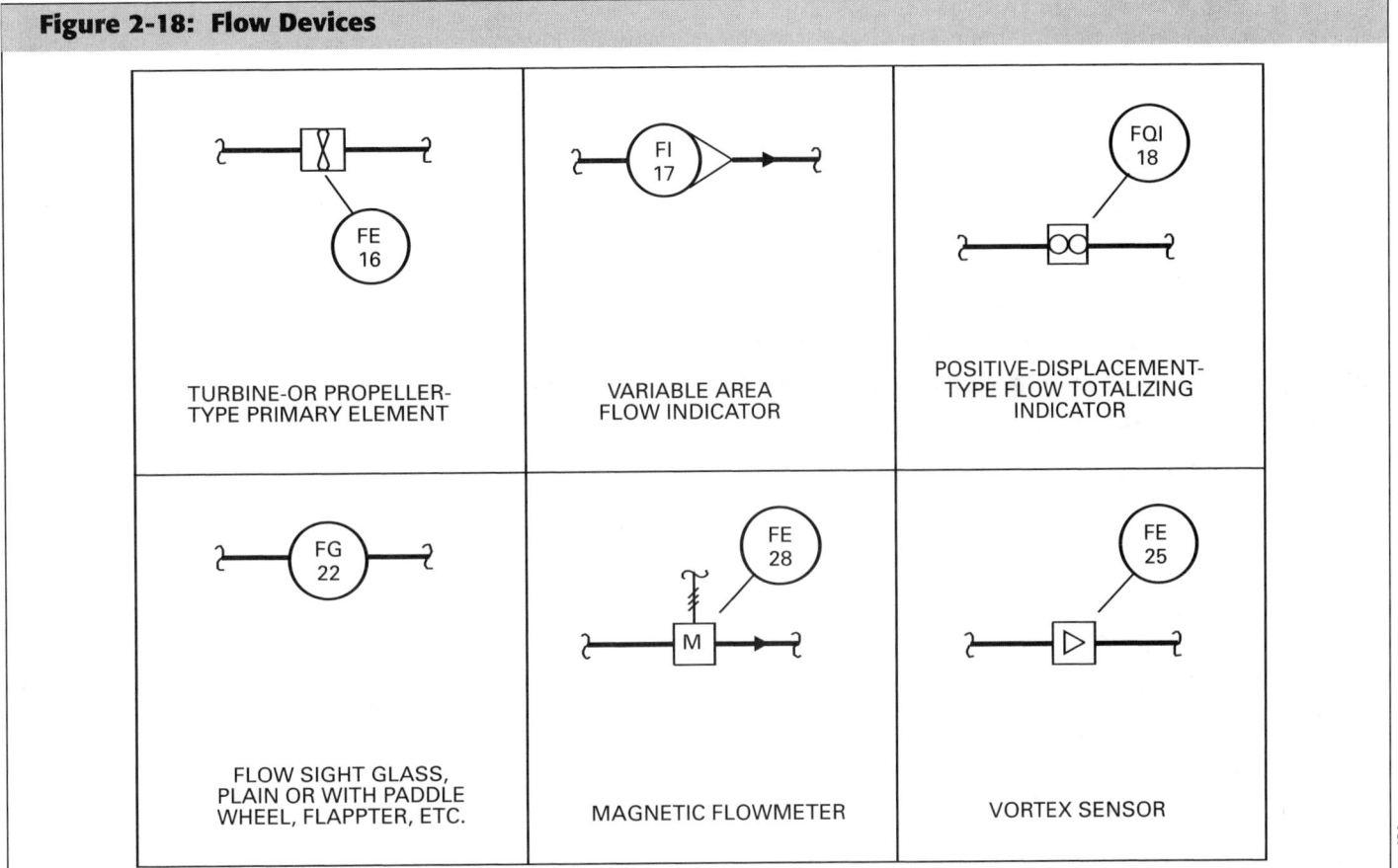

*From ISA-5.1*

Other methods of flow measurement are shown in Figure 2-18. A turbine meter measures the varying speed of a turbine blade in a flow stream. A variable area meter, also known as a rotameter, measures flow through the relative position of a "float" or plummet against a graduated tube. A positive displacement device is used to measure liquid flow volumetrically, such as the water meter in a residence. A sight flow glass is a glass window set into a process line to indicate, but not measure, flow. A magnetic flow meter measures the very small voltage developed when a conductive liquid passes through a magnetic field. The vortex meter measures the change in a process stream as a vortex develops and recedes.

We have presented an overview of the symbols in ISA-5.1. As a review, please do the following exercises.

Match the descriptions from Figure 2-19 with instrument symbols taken from Figure 2-20. When you have finished, check your answers with the answer sheet in Appendix A.

**Figure 2-19: Descriptions**

Instructions: Match the drawing/symbols on the next page with the instrument function title/description below.

1. ( ) Pneumatic Line Symbol
2. ( ) Discrete Instrument - Primary Location Normally Accessible to Operator
3. ( ) Safety Valve
4. ( ) Discrete Instrument - Auxiliary Location Normally Accessible to Operator
5. ( ) Board Mounted Electronic Level Controller
6. ( ) Butterfly Valve
7. ( ) Back Pressure Regulator - Self Contained
8. ( ) Internal System Link - Software or Data Link
9. ( ) Discrete Instrument, Normally Inaccessible (Behind the Panel)
10. ( ) Shared Display or Control - Primary Location Normally Accessible to Operator
11. ( ) Electromagnetic or Sonic Signal, Not Guided
12. ( ) Electric or Electronic Signal
13. ( ) Variable Area Meter (Rotameter)
14. ( ) Control Valve - Pneumatic Actuator, Fail Open
15. ( ) Electric or Electronic Signal
16. ( ) Discrete Instrument - Field Mounted
17. ( ) Control Valve, Fail Closed
18. ( ) Pneumatic Binary Signal
19. ( ) Pressure Indicator
20. ( ) Programmable Logic Controller - Primary Location Normally Accessible to Operator
21. ( ) Flow Gauge

## PFD Defines Process Conditions

As the project design progresses, information from the PFD is used to define process conditions for equipment and piping. The equipment or vessel group sizes vessels using information first established on the PFD. The piping group, or perhaps the process group, calculates the pipe sizes. The mechanical equipment group selects equipment. Equipment requirements may influence the process throughput, which has an iterative impact on the process design. Equipment changes introduce process changes that change line sizes. Equilibrium is reached eventually, as the project team establishes more details and pertinent information becomes available. All this information is recorded and updated on the P&ID. The P&ID is the coordinating document among design groups. Each design group must continually add information to the P&ID and check the information added by other groups. As piping and equipment details become available, the instrumentation and controls design group establishes the process-sensing points, calculates the control valve sizes and begins to add control loop definition.

## Figure 2-20: Symbols

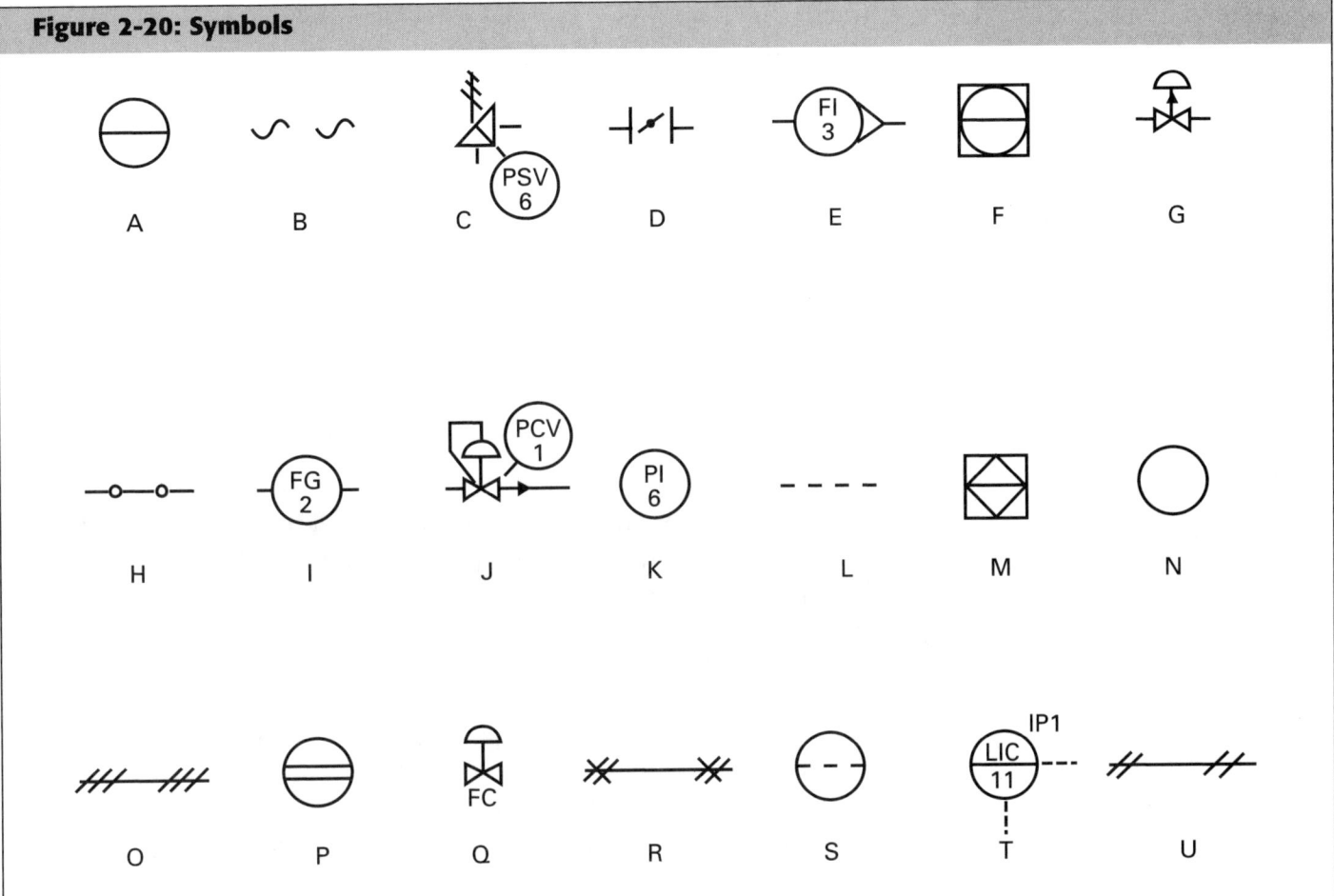

Frequently information added to a P&ID is a coded reference to more complete data that is maintained separately on Specification Forms (data sheets) and databases. The number shown on a pipe is an alphanumeric coded number. The number provides access to additional information on a line list and in the piping specification. This additional information might include materials of construction, pressure ratings, connection methods and service.

## Detailed Design

At some point, the project's decision-makers conclude the P&IDs are sufficiently developed to start the detailed design. The design process goes into high gear. Some projects mark this point as a schedule milestone. The P&IDs, and perhaps other drawings and documents, are formally issued for detailed design. They are so marked on the revision blocks of drawings issued for detailed design, or some similar variation. This point corresponds to a significant ramp up in staffing and the start of design document generation.

The instrumentation and controls design group then increases its activity to place symbols and tag numbers on the P&IDs to depict each instrumentation and control system device or function. Instrument tag numbers indicate the process variable and the device's function. The tag number provides access to more complete information on the Specification Forms and other instrumentation and control systems documents.

There are no universal standards that address the format to be used in developing P&IDs. The format used by most design groups has been developed over many years. However, here are a few guidelines that serve as a simplistic *de facto* standard:

- The process flows from the left of the P&ID to the right.

- P&IDs are developed as "D" size sheets (22" x 34") or larger, but should be legible when reduced to "B" size (11" x 17") for ease of use in the office and in the field.

- P&IDs should show sufficient information to define the process without crowding. One to three pieces of equipment with auxiliaries is normally sufficient for one P&ID.

- To reduce clutter, a typical detail can be used for repeated components (see the "typical drain" on Figure 2-21).

- When piping gets complex, auxiliary P&IDs are used.

- Add notes for understanding and clarity.

- Show relative sizes of equipment, but do not include specific elevations or dimensions.

- Every set of P&IDs should include a legend sheet, or sheets, to define the symbols and abbreviations used.

- The free space on a P&ID should facilitate addition of future process changes; it is best not to start with congested P&IDs.

There is always a trade-off on the subject of how much information and how much detail should be included on P&IDs. Most of the specialist groups tend to want more information on the P&IDs, and project groups want to show less information and keep P&IDs uncluttered. Therefore, it is also good to ask yourself, "Is this information really of value to the end users of the P&ID? Or, is this information kept better elsewhere?"

For this book, our project team uses the Process Flow Diagram from Chapter 1, Figure 1-1, to develop a P&ID, Figure 2-21. The P&ID includes the KO Drum, 01-D-001, and its associated equipment, piping and instrumentation and control components.

Figure 2-21 includes examples of several control loops. This example P&ID also includes several different methods of documenting the control components. It shows how information might be displayed on a P&ID, but it is not meant to be a realistic design for a KO Drum and associated equipment.

**Figure 2-21: P&ID**

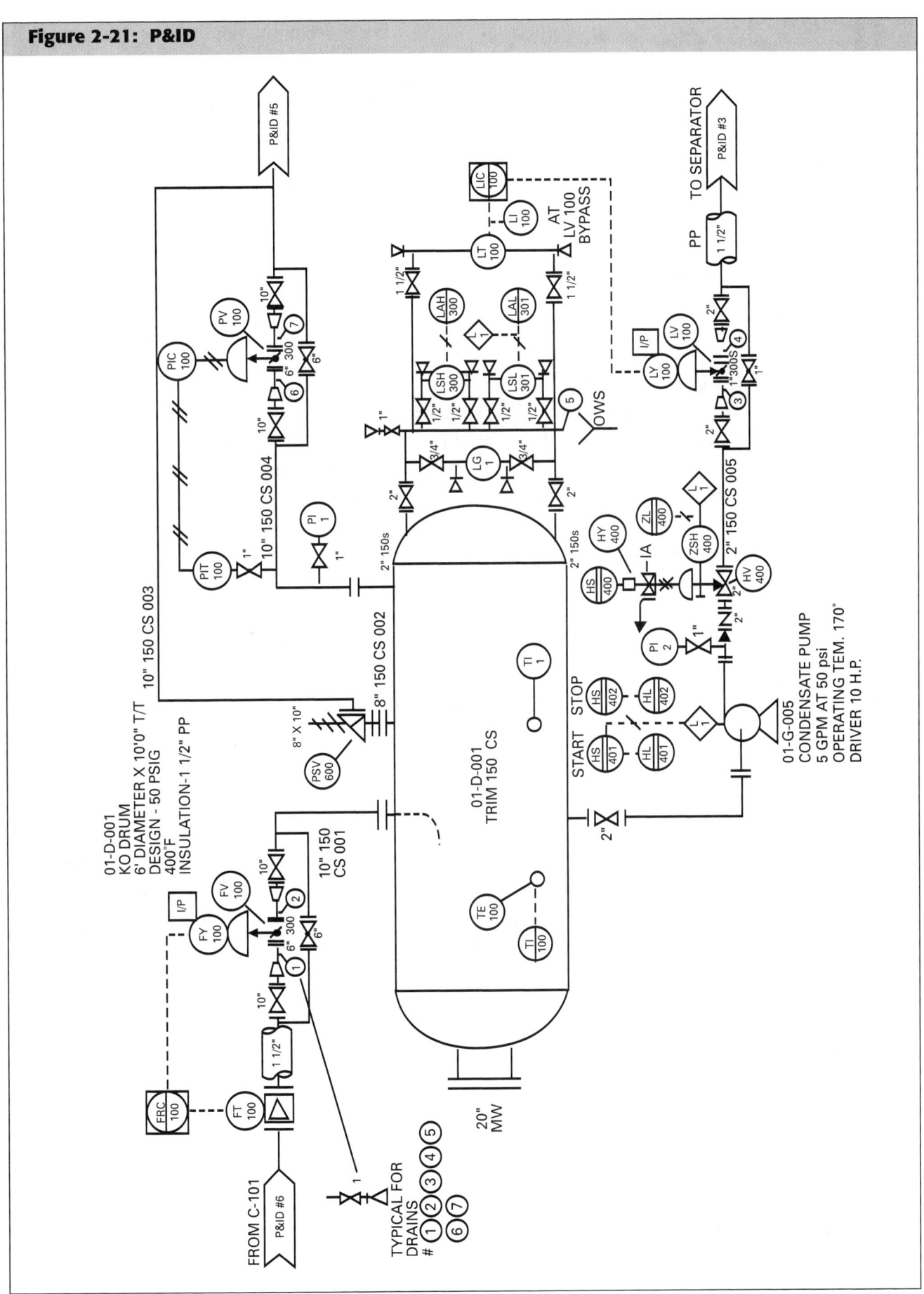

## Equipment Identification

The drawings developed during design will be identified, at a minimum, by a drawing number and a revision number or letter. This information, and more, will be included on a title block. Title blocks have not been shown on the P&ID for simplicity and to conserve space. We will address drawing numbers, revision numbers, and letters and title blocks in Chapter 9.

A unique number is used to identity equipment on P&IDs. For example, equipment number 01-D-001 appears in at least two places – on the PFD and on the P&ID. The D-001 used on the PFD has now been expanded to include a prefix, 01, to signify the drum is located in plant 01. In this example, the rating for the trim piping, ANSI 150 carbon steel, is shown directly on the vessel symbol, 150CS. Additional information is frequently shown on the top and the bottom of the P&ID drawing. In particular, equipment rating, size and nominal or design throughput is provided on the P&ID. However, there is no standard way of showing this information. Some designs show equipment information within or adjacent to the symbol itself. Others will show the detailed information along the bottom of the drawing, relying on proximity to tie the symbol and the data. Or, possibly the equipment number is used to link the data and the symbol.

The vessel group has designed vessel 01-D-001 and the P&ID symbol now reflects that design. It is a horizontal vessel, six feet in diameter and ten feet from tangent to tangent. The tangent lines define the cylindrical part of a vessel. The heads complete the vessel. Also shown are a 20″ diameter access port, which used to be called a manway (MW), and the internal piping to direct the incoming wet gas.

As shown in the vessel data text on the drawing, the vessel is designed to withstand a maximum internal pressure of 50 psig at 400°F. There is a lot of information included in the two simple letters PP within the vessel information text. The "1 ½″ PP" means the vessel will be insulated with one and one-half inches of insulation for personnel protection. This insulation is placed on equipment and piping to protect personnel from injury through contact with a hot surface. The entire vessel is not insulated because we want the gas to cool. The portion of 01-D-001 to be insulated is defined in a plant or project specification – perhaps that which can be reached from the ground or platform, or might be touched when climbing a permanently installed ladder.

Our P&ID shows pump 01-G-005 information adjacent to the pump symbol. Other designs may show more or less information elsewhere. We have elected not to show a symbol for the pump driver, although some designs may include this information. We have shown pump driver start and stop information. When every motor in the facility is started and stopped the same way, the motor data can be provided by a single typical symbol, with details furnished for exceptions only.

Piping and instrumentation and control system connections are now sized and shown on the P&ID. These connections require coordination among several groups.

For example, the instrumentation and controls group has added to the P&ID the symbols for a thermocouple (TE-100) and its well (the small circle). The vessel group and the instrumentation and controls group agree, after consulting the project and plant specifications, that the vessel connection, size and type will be a one-inch 3,000 psig threaded coupling. The agreed locations will be included on the vessel design drawing and on the Location Plan. See Chapter 8 for more information on Location Plans.

The vessel design drawing is typically a stylized plan and elevation schematic showing the layout of the vessel with a connection schedule listing the purpose, size, rating, connection information and location of all the connections furnished by the vessel fabricator. Remember that the vessels go out for bid, purchase and fabrication relatively early in the project, so the control systems design team must focus on defining the vessel requirements early to support that schedule.

The piping group has determined the size and rating for the main process piping, based on the information on the PFD, and the project and/or plant specifications. They add this information and line numbers to the P&ID and then flesh out the balance of the secondary piping information as the design progresses. The vessel group uses this information to specify the vessel connection size and type. Our P&ID uses a simple line numbering system to identify all lines. There is no industry standard for line numbers, although PIP may be addressing the issue. Our system includes the line size in inches, the pressure rating of the line in pounds per square inch, the material of the line by an abbreviation, and a sequence number. For example, the incoming line to 01-D-001 is 10" 150 CS 001. This is a 10-inch carbon steel line rated at ANSI Class 150 and identified by sequence number 001.

Since there is no industry standard for line numbers, other designs might show more or less information on a P&ID. Some designs show only a sequential number on the P&ID. All additional information is shown on a separate line list or a separate database. Some line lists show information defining the start and end of the line, "From and To" information. The pipe schedule and nominal pressure rating is normally provided. Design flow rates can be shown in the line list, but this information might be better provided, maintained, and coordinated using the P&ID. On our P&ID we have shown the nominal pressure rating.

Other designs include more complex line numbers on the P&ID. Many include a symbol or an abbreviation for the service, which calls out the material flowing in the line. A sampling of these abbreviations includes:

| | | |
|---|---|---|
| A - air | FO - fuel oil | S-25 - 25 psig steam |
| C - condensate | IA - instrument air | S-100 - 100 psig steam |
| CW - cooling water | N - nitrogen | PA - plant air |
| FG - fuel gas | S - steam | PW - potable water |

The line list or data base might include additional information about the flowing material – for example, flow in gallons per minute, pounds per hour, cubic feet per minute, pressure, temperature, viscosity, density, and specific gravity. Standard units should be used throughout the P&ID set. For instance, liquids may always be given in gallons per minute, steam in pounds per hour, air in standard cubic feet per minute, and gases in whatever units are typical for your industry. The paper industry may give units in bone dry tons of wood fiber per day. The units are probably metric if the parent company is European or Canadian. To minimize the space needed for line identification, units are frequently not listed on the line callouts; they are only shown on the legend.

Concurrent with the foregoing P&ID development, the instrumentation and controls group will decide on the overall control scheme. This usually occurs after consultation with the owners' representative and the project process group. Both of these groups have valuable information on the best means of control for a process.

On our example project, the instrumentation and controls group has added three control loops to the P&ID. There is a flow loop (FRC-100) on line number 10" 150 CS 001.

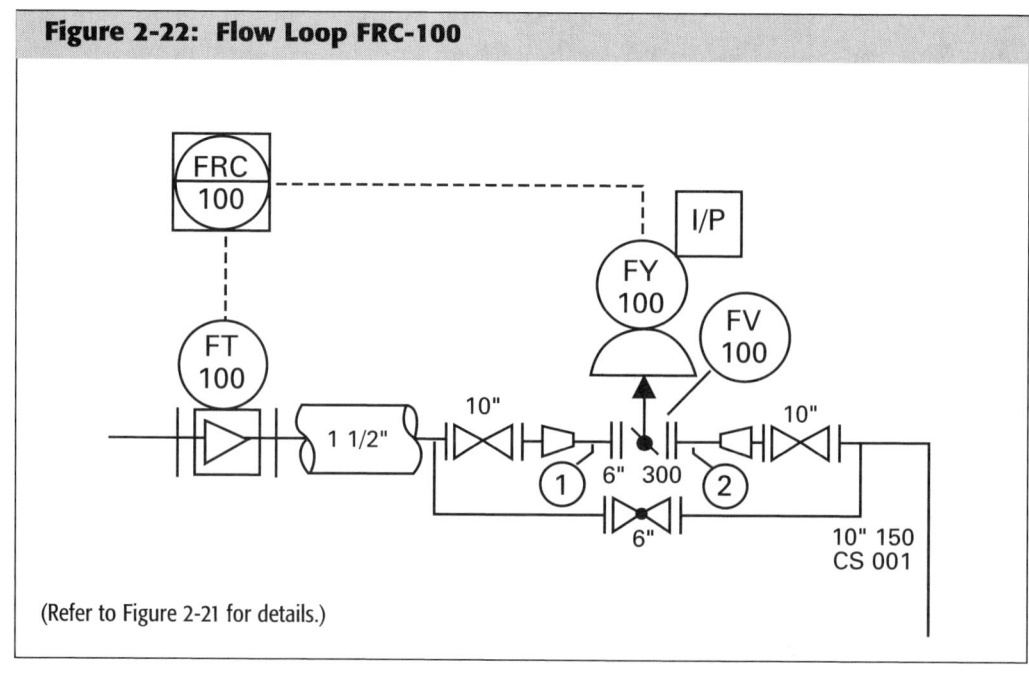

**Figure 2-22: Flow Loop FRC-100**

(Refer to Figure 2-21 for details.)

In Figure 2-22, the electronic flow transmitter (FT-100) is a vortex shedding type. It measures flow by the change in the downstream vortex of the flowing fluid. The vortex is introduced into the fluid by a bar across the flowing stream. The controller (FIC-100) is part of a shared display shared control system – a DCS in our facility. The DCS is located in the central control room or, as stated in ISA-5.1, a "primary location normally accessible to operator".[9]

The signals in the loop are transmitted electronically as shown by the dotted lines. The electronic signal is converted to a pneumatic one at FY-100, the I/P (current-to-pneumatic) converter installed on the control valve, FV-100. FV-100 is a butterfly control valve with a diaphragm operator which includes a spring to open the valve. If the air supply to the valve is lost, the valve will fail to the open position, as shown by the upward arrow on the valve operator. FV-100 is installed with block and bypass valves, as defined, hopefully, in the owner's design requirements' specification. The block and bypass valves permit the control valve to be isolated and process flow continued by manually operating the bypass valve. The instrumentation and controls group has calculated the size of the control valve using the flow, pressure drop and temperature agreed upon with the piping and process groups. The control valve size is 6" and its flanges are rated at ANSI 300 in accordance with the project specifications. The piping group has, therefore, shown reducers in the line to and from FV-100 while using full-line-size (10") shut-off valves.

The instrumentation and controls group will size and document all the relevant control valve information on Specification Forms or data sheets. They send this information to the purchasing group which purchases the valves. The selected valve vendor will supply dimensional information for each valve to the instrumentation and controls group. This information is forwarded to the piping group so they may complete the piping design.

**Figure 2-23: Pressure Loop PIC-100**

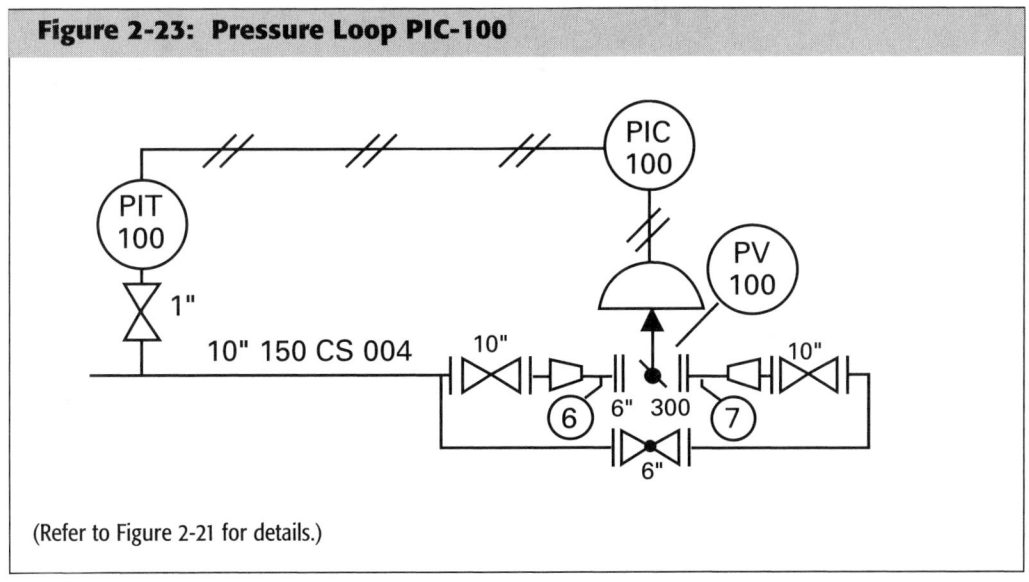

(Refer to Figure 2-21 for details.)

There is a pneumatic loop, PIC-100, on line 10" 150 CS 004. This loop controls the pressure in the KO drum. PIT-100 senses the pressure in the line and transmits a pneumatic signal to PIC-100, a field mounted pneumatic controller. PIC-100 develops the correcting signal and transmits it to PV-100, a 6" butterfly valve. PV-100 fails to the open position upon loss of its air supply.

Figure 2-24, Level Loop LIC-100, is an electronic level loop consisting of LT-100, LIC-100, LI-100, LY-100 and LV-100.

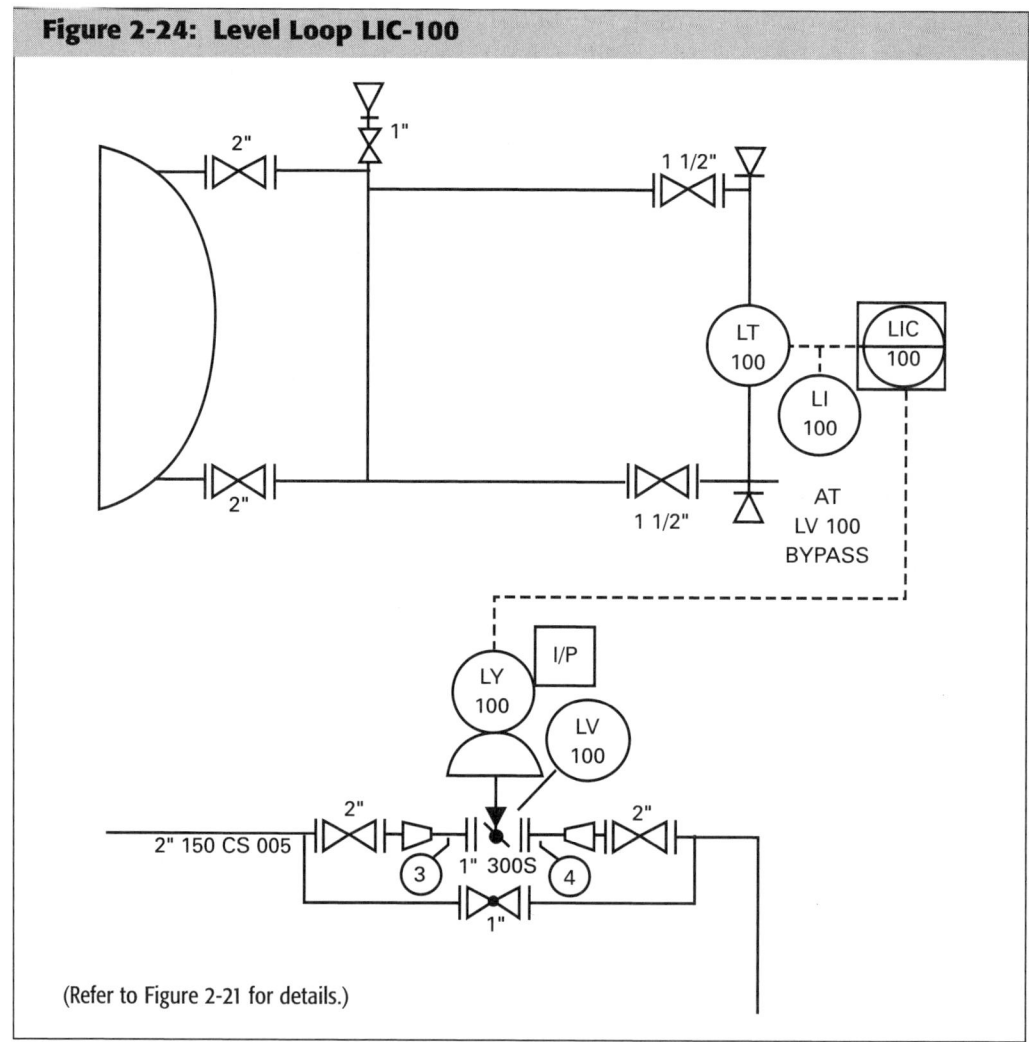

**Figure 2-24: Level Loop LIC-100**

(Refer to Figure 2-21 for details.)

LT is shown as a displacement type transmitter, LIC-100. The indicating controller is part of DCS system. LI-100, a local electronic level indicator is shown by a note to be located at the bypass of control valve LV-100. This might be necessary if control valve LV-100 is out of service for maintenance but the electronics of the DCS are still in service. A process operator could monitor the drum level by manually positioning the bypass valve while watching the drum level on LI-100. LY-100 is the I/P converter. LV-100 is a butterfly control valve that fails closed upon loss of its air signal, since the power to the valve actuator is only the instrument signal.

The piping group added 1" plugged drain valves upstream and downstream of all control valves to drain any residual liquid in the line when valve maintenance is necessary. In use, the drains are connected to a portable recovery system. After the liquid is drained, the control valve may be removed without atmospheric contamination.

The level devices are all connected to a strongback, also called a pipe stand, or bridle connected to KO Drum 01-D-001. A strongback is a continuous pipe, typically 2" nominal, connected to a vessel through top and bottom block valves. The level instruments are mounted to the strongback and include the level gauge (LG-1), high level switch (LSH-300), low level switch (LSL-301), and level transmitter (LT-100).

Level devices are sometimes, but not always, connected to vessels in this manner for several reasons. First, attaching level instruments to an isolatable strongback, or instrument bridle, facilitates testing the instruments without disturbing production, or without opening the entire vessel. If suitable for the process fluid, the strongback can be isolated, the vent valve opened and a compatible process fluid can be introduced through the drain to prove the function of the level devices. Secondly, vessels usually have long delivery times and, therefore, they are purchased early in the project. The level instruments are normally purchased much later. Accurate dimensional information for installing the level devices may not be available when it's necessary to install the connections directly on the vessel. Therefore, in this example, two 2" connections are placed on the vessel for the strongback. The instrument connections to the strongback are scheduled much later in the project.

Thirdly, vessel connections are more expensive than piping connections. We are not advocating connecting all level devices to strongbacks rather than directly to the vessel, but we are showing this type of installation as a possibility.

In Figure 2-25, the buttons and lights (HS/HL 401&402) start, stop, and indicate the state of pump 01-G-005. The tag marks are located in a circle with two parallel lines. This tells us these buttons and lights are located on a local panel or, as stated in ISA-5.1, "auxiliary location normally accessible to operator".[10] This local panel might also be the motor control center. An identifier can be added to indicate which panel contains the components. Usually a three-letter acronym is sufficient to call out the panels. These acronyms must be identified on the legend sheet.

HS-400, also on the local panel, energizes solenoid valve HY-400. This opens HV-400, a pneumatically operated on-off valve. The tee symbol (T) on the actuator shows there is a manual operator on the on-off valve. A manual operator permits the operator to override the pneumatic signal and close the valve.

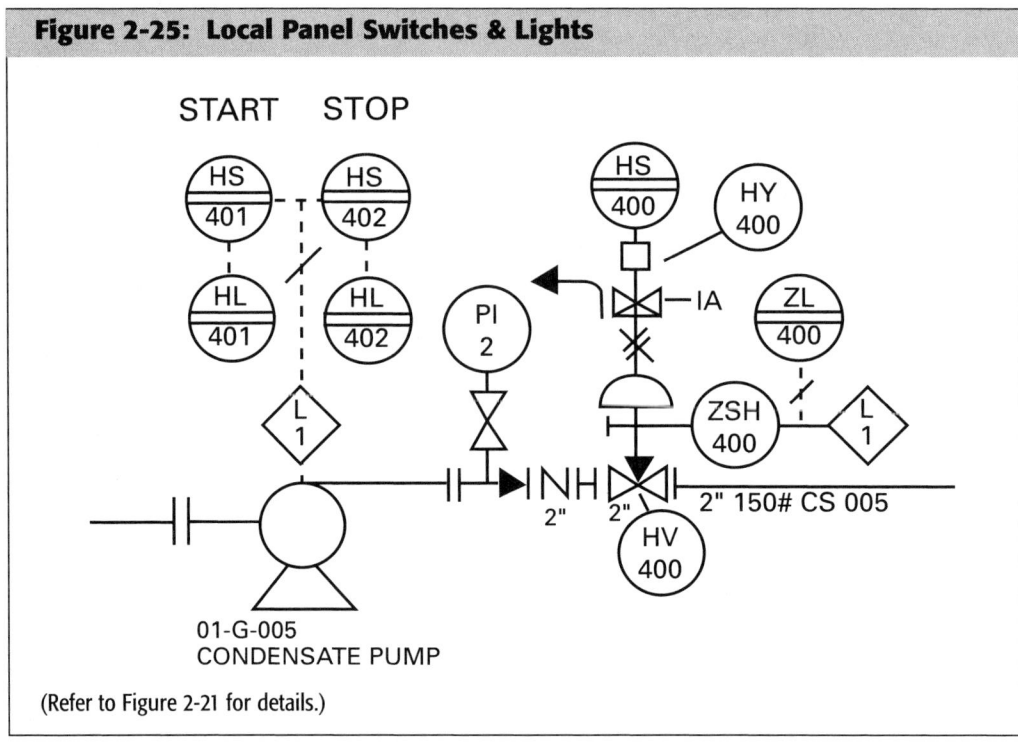

**Figure 2-25: Local Panel Switches & Lights**

As shown by the diamonds with the internal L1 legend, there is some on-off, or logic, or interlock control relating to 01-D-001. The P&ID shows the on-off control involves hand switches (HS 401/402), position switch high (ZSH-401), and level switch low (LSL-301). It is necessary to refer to Logic Diagram L-1 in Chapter 6 to define the on-off control.

The safety valve (PSV-600) has been sized, and a specification form prepared by a member of the project team. Safety valves are sometimes the responsibility of the instrumentation and controls group, sometimes of the process group, or sometimes of the mechanical group. The responsible group will insure (or assure) that safety valves are shown correctly on the P&ID.

There are two pressure gauges shown on the P&ID: PI-1 and PI-2 — one to indicate the pressure in the vessel and the other to indicate the discharge pressure of pump 01-G-005.

To minimize contamination, the strongback drains into an OWS, an oily water sewer. This is a separate underground piping system that is connected to an oil recovery system. A symbol for an oily water sewer is shown as the "Y" with the OWS label.

A typical sketch is shown for the seven drain valves, upstream and downstream of control valves FV-100, LV-100 and PV-100, plus one to drain the strongback.

In this chapter, we have looked at P&IDs in depth, describing what information might be shown on a P&ID and what form that information might take.

We looked at instrumentation symbols and what they mean. We based the instrument symbols on ISA-5.1.

1. *The Automation, Systems and Instrumentation Dictionary*, 4th edition (Research Triangle Park, NC: ISA – The Instrumentation, Systems, and Automation Society, 2000) p.273.

2. ISA-5.1-1984, Instrumentation Symbols and Identification (Research Triangle Park, NC: ISA – The Instrumentation, Systems, and Automation Society, 1984).

3. ibid. p.16

4. ibid. p.21

5. ibid. p.19

6. ibid. p.22

7. ibid. p.28

8. ibid. p.32

9. ibid. p.29

10. ibid. p.29

CHAPTER THREE

# Lists and Databases

Since the 1960s, the lists used for process control work have changed in ways that reflect changes in process control brought about by the application of computational power. Today, in a modern, fully networked process plant, Instrument Lists may exist in electronic form in a database containing all control system documentation. The medium has indeed become the message, with apologies to Marshall McLuhan and his book of the same title.

Instrument Lists used many years ago were compilations of data initiated for some other task. The data was then manually copied to the Instrument List, using the now extinct manual typewriter or an instrument and controls drafter with legible "lettering" skills. The information was typically taken off P&IDs and re-entered on a pre-printed form. In the maintenance shop, calibration cards were scratched out using copies of purchase orders or, over at the DCS configuration desk, I/O was assigned using a smudged termination list and then transferred again to a preliminary Loop Sheet. This manual transfer of information took time and introduced errors, but it was the most practical method, given the available tools.

Indeed, some of us continue to maintain instrument data the same way, albeit with some technology assistance. Possibly, we have gravitated from the old manual typewriter to a computer-based spreadsheet program or even a database. This system may work well for many facilities, but there are still opportunities for improvements in efficiency and accuracy.

---

**Figure 3-1: Instrument List**

- The **instrument list**, or **instrument index**, is an alphanumeric listing of all the devices or functions in a control system together with various drawing and other references

---

Figure 3-1 Instrument List is a brief description of a "traditional" Instrument List or index.

Figure 3-2, A Typical Instrument List, depicts a small part of an Instrument List for a typical project. The list includes the tag numbers for level gauges (LG) and level transmitters (LT) on P&IDs 1, 2, and 3 and references the relevant P&ID, Specification Form, Requisition, Location Plan, Installation Detail and piping drawings. This is a traditional Instrument Index, since the primary function is to show the different appearances of that device in the facility drawing set, much like an index in a book.

**Figure 3-2: A Typical Instrument List**

| Tag # | Desc. | P&ID # | Spec Form # | REQ # | Location Plan # | Install. Detail | Piping Drawing |
|-------|-------|--------|-------------|-------|-----------------|-----------------|----------------|
| LG-1 | D-001-K.O. Drum | 1 | L-1 | L-1 | — | — | ISO-010 |
| LG-2 | D-001 Distil. Column | 2 | L-1 | L-1 | — | — | ISO-015 |
| LG-3 | C-002 Stripper | 3 | L-1 | L-1 | — | — | ISO-016 |
| LT-100 | D-001 K.O. Drum | 1 | L-100 | T-1 | LP-1 | ID-001 | ISO-010 |
| LI-100 | D-001 K.O. Drum | 1 | I-100 | I-1 | LP-1 | ID-002 | — |
| LT-101 | C-001 Distil. Column | 2 | L-100 | T-1 | LP-4 | ID-001 | ISO-015 |
| LT-102 | C-002 Stripper | 3 | L-100 | T-1 | LP-5 | ID-001 | ISO-016 |

## Instrument Databases Boost Efficiency

Today, the information developed and maintained for process control tasks can reside in a more flexible form called an instrument database. The Instrument List and the instrument index are simply a subset of the instrument information available in the database.

Today's instrument databases have the capability to be the primary repository and source for information, rather than a copy of data generated elsewhere. Data can flow out from a single entry in a database to all other documents, rather than from the documents into the database. Our programs can link directly to the database so there are no transposition errors. Some software will also identify changes to notify us when there is a difference between the source data, the database, the application, and possibly the DCS configuration program. For example, if a suffix was added to a control valve after a new control valve was added to a loop, software can immediately, efficiently and accurately compare the source data to the application. The goal is always to write (and check!) the primary source data once, then use it many times.

Instrument databases contain much of the information referenced by different software packages to produce maintenance schedules, calibration records, Instrument Lists, Loop Diagrams, P&ID text and configuration files for the process control computer.

Prior to the explosion of computer usage, we manually typed or wrote listings of the many devices to be purchased and installed in some simple, chronological order. Our field instruments were mechanical or pneumatic in nature, with little electronic computer capability.

When mainframe computers first became accessible, we key punched data cards to describe our instruments and, after some programming effort and some time delay for shipping off the cards and for the return of "computer runs" from the computer center, we were able to develop more useful, specialized lists to simplify our activities.

Engineering and construction contractors developed some of the earliest applications of Instrument Lists. The nature of their work required tracking the progress of the instrumentation and control portion of the project, the specification, bid, purchase and delivery of the instruments. Programs were later written to allow construction progress reporting. Sharing these lists beyond the company that prepared them was difficult, since different engineering, construction or operating companies frequently used proprietary programs and dissimilar software. During this embryonic computer era, our instruments were increasingly self-capable, stand-alone devices that were a bit more flexible but could not communicate very well with other devices.

With desktop computers, even non-programmers can prepare and manipulate data with relative ease. Simultaneously, as instruments became "smart" or "smarter" or "intelligent", the information needed to fully define them increased. Luckily, the desktop computer and database, or spreadsheet software was readily available to manage the additional data. As instruments became more complex and capable, the tools to manage them became more capable as well.

For Instrument Lists, in particular, the single most important technology development may well be the relational database. ISA's dictionary defines it as, "In data processing, an information base that can draw data from another information base situated outside it".[1] In practice, the database is a set of information tables that allows data to be used in an infinite variety of ways, without re-entering the data. Furthermore, the ability to share data among different software packages continues to improve, so information in the database can be used by the myriad of software packages engineering and maintenance use daily.

It is important to remember this feature of an instrument database: to make the best use of the powerful capabilities of the software, information should only be entered once. Once it is entered, it can be used in many different presentations, indexes or reports designed to suit the requirements of the user. For example, the maintenance technician will want calibration reports or component replacement information, the person doing the control system configuration may use function identifiers and I/O assignment data, and the purchasing department may use specification information and status.

## I&C Group Manages Device Documents

The instrumentation and controls group handles information about many individual devices. It is possible this group handles more unique pieces of information and equipment than any other discipline in an industrial process, with the exception of business departments. The discrete bits of information are not particularly complex, but the devices are numerous and information about them appears in many different documents with many different uses.

Consider for a moment a simple instrument tag number – our ubiquitous pressure transmitter PT-100. The device appears first on a P&ID, then later on a Loop Diagram. Its tag number will also appear on Specification Forms and in purchasing documents, a request for quotation, or a purchase order. The plant instrument shop will maintain a log for the PT-100 – listing, for instance, the calibration range, the manufacturer and the model number. The tag number will appear on the Location Plan. There may be an I/O loading document that defines the termination point for the instrument signal. The construction status of the instrument was probably recorded and updated daily during initial installation of the instrumentation and control system. An Installation Detail will be assigned to or developed for each instrument, to show how the device is to be installed.

Suffice to say there is a lot of data associated with that single device. There may be several lists and indexes to record and present this data. A modern facility can assemble all this information in one place, hopefully in an instrument database. Figure 3-3 shows the complex array of information controlled by one company's instrument database program.

As mentioned earlier, in its most basic form, instrument data was presented as a simple list: "Here are the instruments I care about", with little additional information or sorting. The order was sometimes chronological, so its usefulness may have been limited to assuring all the devices had been specified, purchased or installed. Finding a specific device meant you had to scan for it or know when it was listed. The minimal list approach is used today, often with rudimentary sorting, most often as a checklist. It offers little information to interest the maintenance group or as a design tool.

With the addition of a bit more information, the function of the list is expanded to meet other needs. Fields containing the Specification Form number, procurement information, installation and commissioning milestones provide an effective support for the construction effort by summarizing the status of the instrument installation portion of a project.

The most common form of instrument data lists arose with the addition of drawing references – the instrument index. In fact, the terms are really used interchangeably: list, index and, in today's computer based work, the instrument database. The index provided a cross reference from the tag number to the drawing numbers for the P&ID, the Loop Diagram, the Location Drawing, and all other relevant documents.

**Figure 3-3: Typical Example of a Company's I&C Data Flow Diagram**

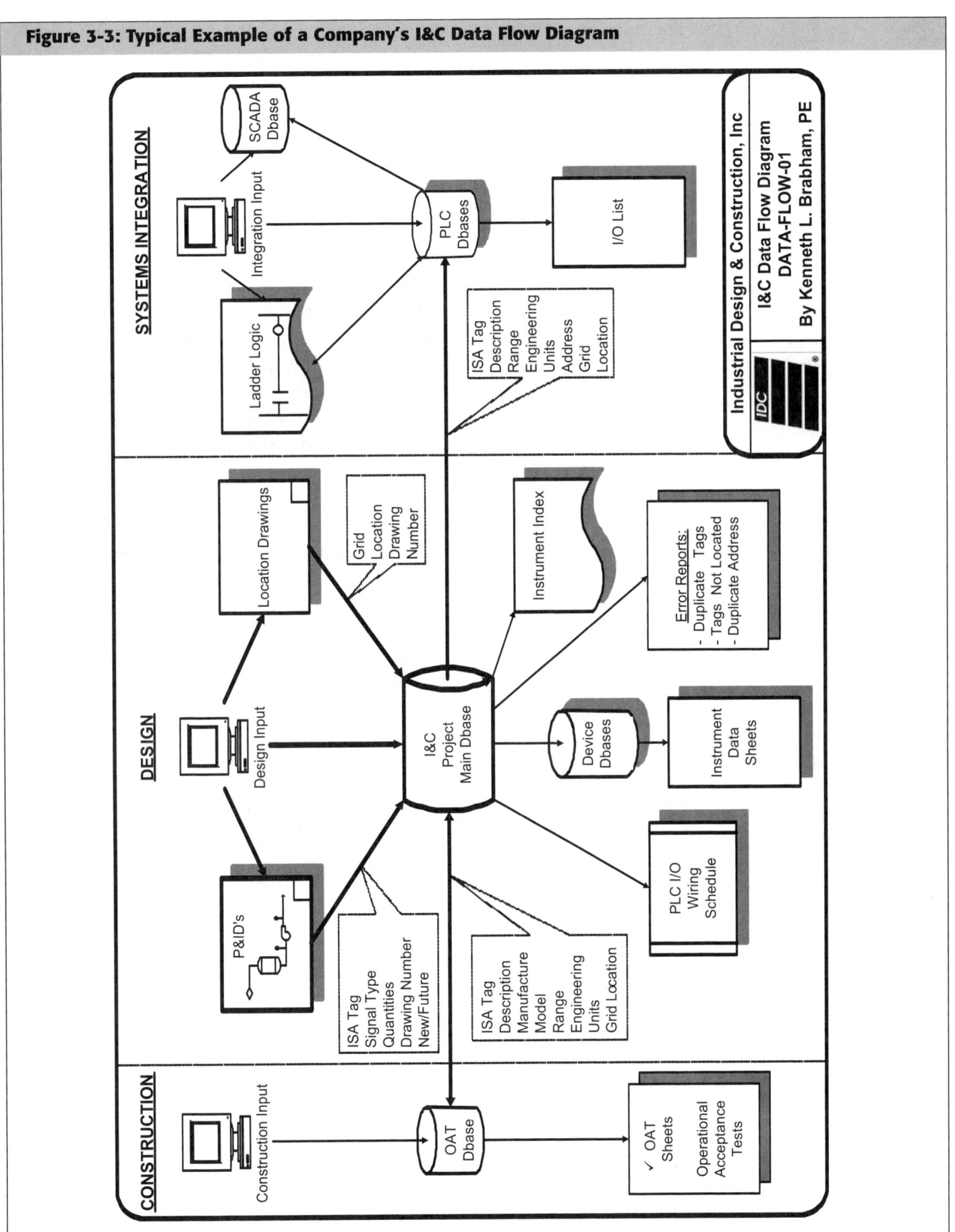

The instrument database in its most complex form contains information limited only by the imagination of the people that use the information, and their budget.

### To List or Not to List: That is the Question

To be useful, an instrument database should list all the tag-marked physical devices that comprise the control system. The general rule of thumb is if something has to be purchased, mounted, wired or tubed, then it should appear in the instrument database. The corollary to this rule, by the way, is that these devices should also appear on the P&IDs, since the P&ID is the source document for the database. The untagged control room operator stations, under these criteria, would then not appear in the Instrument List. However, some, or maybe even most, companies will make up a tag number, specification sheet and database listing for these as well. If listing general components like operator stations suits your purpose, go ahead and list them. Remember to always be consistent in listing equipment. If you tag and list one operator station, you should list them all.

Deciding what information to include in an Instrument List can actually be contentious at times. For example, spirited discussions can take place within process control teams as they decide if software functions are to be included in the instrument database and Instrument List. It is useful then to establish criteria for information maintained within the database:

Is there a need for the information? Don't list data just because it is available. It has to be needed by someone on the document's distribution list.

Is there value added to have this information? Listing each data record has an associated cost. The software has to be modified to carry the information. The data has to be found, entered into the database, printed and checked. Once entered, data in the database has to be maintained, so be sure to consider the cost of maintaining the data. Having out-of-date or inaccurate data is WORSE than having no data. If you have no entry in the database, then someone will go look elsewhere for the information and will probably find it to be both current and accurate.

Enter data ONCE and only once. If the same information is entered twice, entry errors can cause the information to be considered wrong in both places!

The data and database should be used! Encourage widespread expansion of reports and data linking. If no one uses the database, there was no value making and maintaining it.

You have to have a system to maintain the data, a method to track changes to the data and you should regularly distribute the reports. The key word is to have a system, an agreement of procedures, and you should have someone responsible in charge of the database. Anarchy and databases don't mix.

## An Instrument Database Success Story

Figure 3-3 shows one company's instrument database system. It is a complex one that was developed over several years to meet ongoing challenges – and it has proven to be very successful in efficiently maintaining information accurately. One of the primary developers of the system, Ken Brabham, PE, offers some insights into its development and application.

"The necessity for our database architecture arose for the following reasons: to reduce the drawing count, minimize data entry time, to improve accuracy and to complete projects in less time. This approach has now been used on many projects. A typical control system for this database consists of distributed PLCs with a centralized SCADA system. Project size is normally in the range of 3,000 to 5,000 real world I/O points.

"The database was originally developed to handle design information – primarily to generate construction documents like data sheets and wire schedules. The schedules were written and presented in a format designed to replace manually drafted wiring diagrams, thereby eliminating hundreds of I/O drawings for each project. The database is linked directly to the P&IDs so when a drafter adds or modifies an instrument tag on a CADD P&ID, the system includes software to automatically reflect the change in the database. The system became more than just a database of information – (it became) a design tool for all projects.

"This database not only became an invaluable tool for the design phase of our work, but as projects transitioned from design to the systems integration phase, we found many additional uses as well. One significant use stems from my inability to type at a measurable rate! My slow keypunch speed led me to hone my database programming skills, so I now hardly need to type at all. Since the PLC and SCADA software packages support data importing and exporting, all that's needed is to simply reformat the design data so it can be directly imported into the various control systems. This way, the PLC programmer never has to enter any of the real world I/O information. He or she simply imports the information as needed: the instrument tag number, point description, I/O assignment and ranges for analogs. This greatly reduces the amount of time spent digging through drawings and specifications to obtain such information.

"When the PLC programmer completes programming on a system, the data is exported and once again reformatted for importation into the SCADA package. However, this time all SCADA points can be imported, not just the real world I/O points.

"Now both the PLC programs and the SCADA points have been generated from the design database without the need to retype tags, descriptions and addresses. In fact, the instrument tags had only been typed once by the drafter during the P&ID development phase, and that information was automatically extracted and imported into the database. Since all of this was obtained directly from the design database, human errors are greatly reduced. We also discovered that once you maintain this information in a database, you continue finding additional benefits like error reporting for duplicate tags or duplicate I/O assignments. We have even used the database system to track individual instrument status from procurement through installation and acceptance.

"What's the downside to a database approach, you ask? Well, for me, I still can't type." – Ken Brabham, PE[2]

The discussion regarding the need to list software functions (Control, Ratio Control, Feed Forward, Sum, Integrator, Record, Trend, Bias, etc.) in the database would probably consider whether or not a master list would be of use to the people preparing the control computer configuration. For instance, the software functions might be an important review and approval item between a design company and the client, so there could be advantages to including the functions in the database. Listing them could simplify the approval and progress report development. This is an example for discussion because listing software functions in the instrument database or Instrument List is not a universal practice. Commonly used fields of information in an instrument database are listed in Figure 3-4.

**Figure 3-4: Instrument Data Fields**

**Basic Data**
Tag Number
Function or Service
I/O Type (AI, AO, DI, DO)

**Technical Data**
Calibration Range and Units
Rating
Power: Loop or 120V AC

**Index Data**
P&ID
Loop Sheet Number
Installation Plan Number

**Connection Data**
Junction Box
Marshalling Panel
I/O Rack
Address

**Maintenance and Operations**
Stores or Stocking Number
Calibration Date
Calibration By
Manufacturer
Model Number
Vendor
Vendor Phone Number
Purchase Order
Original Delivery Date
Current Delivery Date
Received – Yes/No
Receiving Report Number

**Construction Data**
Calibration or Shop Approved
Issued to Contractor – Who?
Issue Date
Installed Date
Tubing Complete - Date
Wiring Complete - Date
Checkout Date
Commissioning Date
Turnover Date

The data fields listed in Figure 3-4 are only suggestions. Fields that you use should be selected to meet your facilities requirements. As mentioned in the criteria above, there should be some value returned to you and your company for maintaining the information. If no one at your facility cares if a control valve has an open or closed yoke, then there is little need to maintain that information in your database. The more useful the database is to you, the better the information it will contain.

Construction information may appear in a database, as shown in Figure 3-4. In fact, procurement or construction activities are likely to initiate the instrument database. The maintenance database used five years from now probably will have started as a tool to monitor construction progress on a large project. Yet, after construction is complete, this information will probably never be referred to again. In a relational database, construction information can be maintained in a separate table that could be abandoned after the facility is turned over to operations people. At the same time, a maintenance table of information will come into use. Calibration and maintenance work will be scheduled, performed and recorded in a database for the remainder of the facility's life.

## Database Becomes Master Document

The instrument database, or Instrument List, is, or should be, the master document to record and maintain tag numbers and function description. This data appears in many places including Specification Forms, Loop Diagrams and possibly on your DCS screen. The function description is sometimes referred to as the loop title. It is both professional and much less confusing to have all the text fields read the same wherever they appear. The instrument database ensures this happens. The function identifier field should be written only once, no matter how many times it is used. The relational database allows the data to be pulled from the same field every time it is used.

The loop service text field describes, in words, the function of the loop. Having one master document for this information is critical to prevent duplication of tag numbers and to maintain a pattern for your service descriptions. A fixed pattern will make the text easier to understand, and more efficient to develop. Service descriptions are important from a practical standpoint when trying to locate a specific loop while using the Instrument List, or from a practical standpoint, when thumbing through a pile of Loop Diagrams.

There are essentially two formats used for service identifiers, with myriad variations on those formats. The first is to put the loop function first, followed by a "from" and, if needed, a "to" description. The function is, of course, based on ISA-5.1. For instance: "Level – Tank A". Or, another example: "Flow – Chilled Water to Condenser".

The hyphen is just a delimiter for clarity, but the delimiter and its surrounding spaces should be established at the onset of the index, since the delimiters and spaces can impact electronic sorting. Also, if esthetics is important, using the same delimiter and spacing looks neater.

The second way to call out the service is to state what the loop does conversationally. For example "Tank A Level" or "Condenser Chilled Water Flow". The conversational service identifier is more difficult to sort by function, but it is probably more natural in use. It is important that you don't mix methods; once you pick your method, do not allow deviations.

| Figure 3-5: Common Service Description Formats | |
| --- | --- |
| Function Led | Level – Tank A |
| Conversational | Tank A Level |
| Function Led | Flow – Chilled Water to Condenser |
| Conversational | Condenser Chilled Water Flow |

## Ensure Agreement

This may seem like a minor issue, but it is important to work with the operations people to ensure equipment titles you use are the ones they are familiar with. Calling a tank the Secondary Hot Water Tank may be perfectly correct, unless everyone for the past 35 years has called it the "Green Tank". When the maintenance I&C technician is called in at 3 a.m. to fix the level loop on the "Green Tank", he or she won't really care that your title was "more correct". However, they will care that they had to search around for 20 minutes to find the right troubleshooting documents. Hopefully, they will share their experiences with you.

When you are part of a design project, it is imperative that the equipment titles and equipment numbers used in the instrument database are the same as those used by the equipment and electrical groups. This is the power of the relational database. The instrument database can link, or relate, to the master equipment list. Titles used in the instrument database should agree with those used on the rest of the project.

When developing tag numbers and function descriptions, it is good practice to check the format requirements of *all* the control and monitoring software that use the information. Establish the character counts and formats used to make up the text strings. The Instrument List or instrument database can then provide the template for creating the text that ensures the correct use in all the applications. For instance, your DCS may limit character fields for a service description associated with a tag number to 32 characters, so your service text must follow that format. This is easy to set up in an instrument database.

The Instrument List, index or database is a core document prepared by the instrumentation and control group. A good Instrument List is a summary of the physical portion of the instrumentation system. The document has undergone somewhat of a renaissance with the expansion of computers and the inter-operability of software. Many others can use information stored in one software package. With the computer skills contained within today's instrumentation and controls groups, it is only natural that some very useful information-sharing has arisen offering increased efficiency, speed and accuracy for designing and maintaining systems.

Recent projects have linked the P&IDs and the Location Plans to the instrument database. The capability exists to automatically extract the instrument tag number from the P&ID, as well as the associated piping line number, P&ID number and associated equipment number. Indeed, any information that appears on a drawing is potentially available for extraction to other software, assuming the CAD software and the extraction software are compatible.

Since Location Plans are made to scale, it is now possible for software to be written that will search for a tag number and feed back a grid reference location for that device. This both simplifies development of the instrument database and increases the accuracy. The tag and equipment number will be exactly as shown on the drawing, since that is the information source.

If no location is retrieved for an instrument, then that instrument has not yet been located. Or, if no I/O assignment has been returned from the card loading drawings, then that device has not yet been assigned. There are pitfalls, of course. The "information mining" software might require that drawing data be entered in a particular way. It may be necessary to train staff to manipulate the software to enter and extract the information in a useful manner.

## Software Development Challenging

The biggest challenge for any instrument database is developing the software. Third party software is usually available to perform the desired tasks, at some expense. You can write your own software in-house, again at some expense. Third party software allows quick startup of your instrument database without having to learn the language of databases. Third party software may also offer the advantage of continuous technical support independent of your own staffing.

Many companies have spent significant money developing software to perform a specific task, only to have the programmer depart, leaving the software package unsupported. This can happen with third party software also, so care should be taken to pick a reputable, respected company that will be around a year from now. You should get references to call. It may be useful to look at the supplier's software development history to get a sense of how they handle upgrades.

Happily, there is a trend in computer usage that bodes well for using databases for Instrument Lists and reports: we are all becoming much more comfortable with some pretty complex computer software. As more high schools teach software classes, skills to manipulate data in meaningful ways are becoming more universal. A software package that looks impossibly mysterious and daunting to someone who left school before 1980 may look like, well, just another type of a pencil to someone in school right now. As a result, manipulating data in a relational database is becoming simpler as we all become more experienced with available tools.

---

1. *The Automation, Systems and Instrumentation Dictionary*, 4th edition, Research Triangle Park, NC: ISA — The Instrumentation, Systems, and Automation Society, 1984). p.419

2. Brabham, Ken, PE. Unpublished notes

## CHAPTER FOUR

# Specification Forms

When the Instrumentation and Controls Design Group adds device symbols and tag numbers to a P&ID, they simultaneously undertake two other activities. First, they enter the tag numbers on an Instrument List, or in a database, using the list to ensure the tag numbers are unique. Second, they prepare a written definition of the instrumentation and control devices. This definition is most often, but not always, on a Specification Form, sometimes referred to as a data sheet or a spec sheet. Other methods of definition include bills of materials, requisitions, or a simple descriptive paragraph. Items purchased "off the shelf", or bulk material items that do not require specialist knowledge to select, will frequently be called out on one of these alternates – for example, items such as pressure gauge isolation pigtails, condensate traps, and pulsation dampeners.

Clearly, the information available on a P&ID is insufficient to accurately and completely define instrumentation and control devices, so additional information and a way to present it is needed. Referring to Figure 2-21 (page 45), the symbol and tag number define FT-100 as an electronic vortex shedding flow meter. However, this is only part of the required definition. A potential supplier could not accurately quote a price and delivery for the device without additional information. The supplier will need to know, as a minimum, the process fluid to be measured, pressure and temperature rating of the system, allowable piping connection information, power and signal requirements of the device, electrical area classification, and the housing rating. To take full advantage of the supplier's expertise, the supplier will need to be told the minimum, normal and design flow rate, operating pressure and temperature of the process, process fluid information, including viscosity, and metallurgical and material requirements.

Since a great deal of information is required to define each instrumentation and control device, a Specification Form is used. This aids in organizing and assuring complete and consistent information. Blanks on the Specification Form imply missing information.

The need for uniform Specification Forms is stated in ISA-20-1981, Specification Forms for Process Measurement and Control Instruments, Primary Elements, and Control Valves, as follows:

"Because of the complexity of present day instruments and controls, it is desirable to use some type of specification form to list pertinent details for use by all interested parties. General use of these forms by users and manufacturers offers many advantages, as listed below:

1. Assists in preparation of complete specification by listing and providing space for all principal descriptive options.

2. Promotes uniform terminology.

3. Facilitates quoting, purchasing, receiving, accounting, and ordering procedures by uniform display of information.

4. Provides a useful permanent record and means for checking the installation.

5. Improves efficiency from the initial concept to the final installation."[1]

The preceding quote was written specifically for ISA-20, but it makes the case for the use of uniform sets of Specification Forms. Figure 4-1 provides a definition of a Specification Form.

---

**Figure 4-1: Specification Forms**

- **Specification forms**, or **data sheets**, define the tag-marked devices that make up a control system

- The form contains information necessary to secure vendor quotes and to purchase devices

---

Development and management of data needed to complete Specification Forms is a significant part (perhaps 25%) of the instrumentation and controls group work-hour budget.

There are many variations of Specification Forms. Most engineering contractors have developed a set for their use; some instrumentation and control device suppliers have their set, and most of these are based on the ISA-20 standard. ISA also has a newer technical report, ISA-TR20.00.01-2001, Specification Forms for Process Management and Control Instruments-Part 1: General Considerations. (See Appendix C, page 163.) It is also available in Microsoft Word Format on CD-ROM.

Of course, everyone prefers to use the forms with which they are most familiar. A Specification Form modified by a third party typically will present the data in a similar format, but will normally drop fields not applicable for the typical devices used in that industry. The ISA form for pressure transmitters includes sections on controllers and chart recorders that are not applicable to intelligent or smart electronic transmitters, so the form used by some companies will drop those sections and add fields needed for the devices used. A modified Specification Form may include a definition of the communication bus used, plus a place to list the information needed by the device supplier to pre-program smart or intelligent transmitters before shipment.

The intent of all of the forms variations is the same. A Specification Form will define the instrumentation and control devices completely and accurately so vendors may quote on, and supply, the original device and replacement; so construction personnel can install and check out the device; and so maintenance personnel can calibrate and repair the device.

ISA-20 consists of Specification Forms for twenty-six commonly used instrumentation and control devices. We all tend to forget ISA-20 also provides a line-by-line explanation of information intended to be furnished in the fields.

This is an invaluable reference, particularly for those working with Specification Forms for the first time – and they are a pretty good review for those that have been working a while in other areas. ISA-TR20.00.01 is a newer set of Specification Forms issued by ISA as a Technical Report. The new Specification Forms have been modified and expanded from the original ISA-20. There are also entirely new forms for devices not included in ISA-20.

## Specific Knowledge Required

When filling out a Specification Form, the instrumentation and control specialist must have at least three types of knowledge:

1. **Process Knowledge:** The specialist must know how the process "works" and how it can be controlled. The specialist must know the process flows, process pressures, process and ambient temperature and, very importantly, any expected variation in the process conditions. This person must know both the effect of the process fluids on all wetted parts and must understand the effect of exposure to corrosive or explosive materials. The specialist must be aware of, and familiar with, requirements set forth in the local and national codes. In the United States, this is the National Fire Protection Association (NFPA) Standard 70, National Electric Code (NEC).

2. **Specification and Standard Knowledge:** The specialist must know what specifications and standards are to be followed. Many operating entities and engineering contractors have comprehensive specifications that dictate the quality of the materials, equipment and documentation to be supplied for a project. There are also many industry standards that may be relevant and must be followed.

3. **Product Knowledge:** The specialist must also know a great deal about instrumentation and control device suppliers, including which vendors can supply each type of device. When competitive bids are issued for instruments, the specialist will need to know what features are common to all devices, regardless of vendor. It is necessary to recognize what features are "hot selling points" made to seem important by a particular salesperson, but not actually required.

The fact that XYZ Company has a new transmitter with a guaranteed accuracy ten times better than the competition is interesting to know. Understand though, if higher accuracy is made a requirement by listing it on the specification form, it will prevent all but one vendor from supplying the device. One of the tasks of the instrumentation and control specialist is to know what features are *really* required. For a competitively bid project, a specification that limits the number of prospective suppliers needlessly will result in higher costs for the devices. The other side of the argument is a specification that is too loose will result in poor operation of the plant.

## Alternative Means of Definition

Up to this point we have concentrated on the Specification Form as the means of defining instrumentation and control devices. As mentioned earlier in this chapter, other means of definition are also used.

Some projects, or firms, use a coded tag number for some uncomplicated instrumentation and control devices shown on a P&ID. The example in Chapter 2 uses PI-100 to designate a 4-1/2" diameter pressure gauge, with a range of 0-150 psi and a stainless steel Bourdon tube. The PI-100 might appear 50 times on a complete project set of P&IDs. Another pressure gauge tag, PI-101, might also be used for gauges similar to PI-100 except with a different range, 0-200 psi. PI-101 might appear 25 times on our P&IDs. To prevent confusion, some firms will use a letter designator in lieu of the loop number, to ensure these commodity devices are not mistaken for a loop number.

ISA has developed a software version of ISA-20 containing all the forms and instructions on a single CD. It is possible to select a form, fill it out from a computer keyboard and print the finished Specification Form. The software version of ISA-20 has been used to develop the Specification Forms in this chapter.

In addition, the software version of Technical Report ISA-TR 20.00.01 Specification Forms is available on a CD. The Technical Report contains many new and updated versions of the Specification Forms. We have included the Specification Form for Displacer-Type Level Transmitters or Local Controllers as Appendix C. The pick list permits selecting the variables in a Specification Form and printing them on the Specification Form.

Figure 4-2 is the specification form for pressure gauges from ISA-20, and Figure 4-3 is the instruction sheets to help fill out the form.

We will review the method used to fill out the gauge Specification Form by following the instruction sheet.

1. A direct reading gauge is required.

2. Local mounting is required. The pressure gauges are mounted on a piping connection to a process line as shown in an installation detail. (See Chapter 8, Location Plans and Installation Details.)

3. Dial Diameter 4-1/2". Dial will be white.

4. Case material – Manufacturers Standard Plastic. Let the gauge manufacturer specify the case material.

5. Ring – Standard. Let the gauge manufacturer specify the ring that holds the glass in the housing.

**Figure 4-2: Pressure Gauge – Specification Form**

| | PRESSURE GAGES | | | | SHEET | | OF | |
|---|---|---|---|---|---|---|---|---|
| (ISA) | | | | | SPEC. NO. | | REV. | |
| | NO | BY | DATE | REVISION | . 431 | | 0 | |
| | 0 | FAM | 12/15/2003 | | CONTRACT | | DATE | |
| | | | | | 1234 | | | 1/3/2003 |
| | | | | | REQ. - P.O. | | | |
| | | | | | J-1 | | | |
| | | | | | BY | CHK'D | APPR. | |
| | | | | | FAM | CAM | LF | |

1. Type:    Direct Rdg ☑    3-15 lb Receiver ☐
   Other _____
2. Mounting:    Surface ☐    Local ☑    Flush ☐
3. Dial:    Diameter 4/12"    Color WHITE 1/2%
4. Case:    Cast Iron ☐    Aluminum ☐    Phenol ☐
   Other MFG. STD. PLASTIC
5. Ring:    Screwed ☐    Hinged ☐    Slip ☐    Std ☑
   Other _____
6. Blow-Out Protection:    None ☐    Back ☐    Disc ☑
   Solid Front ☐    Other _____
7. Lens:    Glass ☑    Plastic ☐
8. Options:    Sylphon Material    ☐ _____
   Snubber    ☐ _____
   Pressure Limit Valve ☐ _____
   Movement Damping ☐ _____
9. Nominal Accuracy Required    1/2% FULL SCALE

10. Mfr. & Model No. _____
11. Press. Element:    Bourdon ☑    Bellows ☐
    Other _____
12. Element Mtl:    Bronze ☐    Steel ☐    316    SS
    Other _____
13. Socket Mtl:    Bronze ☐    Steel ☐    316    SS
    Other _____
14. Connection-NPT    1/4 in. ☐    1/2 in. ☑    Other _____
    Bottom ☐    Back ☐
15. Movement:    Bronze ☐    SS ☐    Nylon ☐
    Other MFG. STD.
16. Diaphragm Seal:
    Mfg. _____    Type _____
    Wetted Part Mtl. _____    Other Mtl. _____
    Fill Fluid _____
    Process Conn. _____    Gage Conn. _____

| Rev. | Quan. | Tag Number | Range | Operating Pressure | Service |
|---|---|---|---|---|---|
| 0 | 50 | PI-100 | 0-150 psi | various | — |
| 0 | 25 | PI-101 | 0-200 psi | various | — |
| | | | | | |
| | | | | | |
| | | | | | |
| | | | | | |
| | | | | | |
| | | | | | |
| | | | | | |
| | | | | | |
| | | | | | |
| | | | | | |
| | | | | | |
| | | | | | |
| | | | | | |
| | | | | | |
| | | | | | |

NOTES:

© 1981 ISA                                                                ISA FORM S20.41a

From ISA-20

6. Blow Out Protection – This lets a part of the case blow out if the Bourdon tube ruptures. Since the operator will read the gauge from the front, specify a blowout disk in the back of the case.

7. Lens – Glass might break; plastic sometimes becomes cloudy. With a back blowout, glass has been specified.

**Figure 4-3: Pressure Gauge – Instructions**

## 21 Pressure gauges

Instructions for ISA Forms S20.41a and 20.41b

1)    When receiver gauges are specified, the "Range" in the tabulation is the dial range.

2)    Select mounting style.

3)    Specify nominal dial diameter.  Dial assumed white unless otherwise specified.

4)    Select case material.

5)    Specify ring style, or check "STD" if not important.

6)    Specify blow-out protection.  "Back" refers to a blow-out back. "Disc" refers to a blow-out disc located in the back or side of the case.

7)    Specify lens material.

8)    Options:
      Snubber                  Specify type or model number.
      Sylphon Material         If sylphon required, specify material.
      Movement Dampening       Specify if required.

9)    Specify nominal accuracy, such as "±1/2%."

10)   Write in make and model number after selection is made.

11)   Specify element type or write in "MFR.STD."

12)   If stainless steel is required, write in the type; such as "316."

13)   See 12.

14)   Specify connection size and location.

15)   Specify movement or write in "MFR.STD."

16)   If Diaphram Seal is required, fill in specifications.

For convenience, write in psig or other pressure unit at the top of "Range" and "Op. Press" columns, if all are the same.

*From ISA-20*

8.    Options – No options specified.

9.    Nominal accuracy required – +/- 1/2% is a good accuracy and does not keep manufacturers from quoting; also, it keeps the costs low.

10.   MFG and Model # – To be filled in after gauges are purchased.

11.   Pressure element – Bourdon tube is usually used for the ranges under consideration.

12.   Element material – 316 stainless steel has been selected.

13.   Socket MTL – 316 stainless steel.

14. Connection – 1/2 inch.

15. Movement – MFG-STD; let the manufacturer specify.

16. Diaphragm – None required.

Then we complete the form as follows:

| Quantity | Tag | Range | Operating Pressure | Service |
|----------|--------|----------|----------|---------|
| 50 | PI-100 | 0-150 psi | Various | – |
| 25 | PI-101 | 0-200 psi | Various | – |

Then fill out the top right portion of the form – the revision number, sheet number, specification number, and the contract number. You are now finished with the Specification Form.

As mentioned earlier in this chapter, some projects do not use Specification Forms to define instrumentation and control devices. These projects might define the devices using bills of material, requisitions, or a simple descriptive paragraph. These methods are not significantly different from using Specification Forms, since similar information is needed for all – complete definition of the device is necessary.

The difference in using Specification Forms or prose is in the handling of the information. Today, there is a real need to have data in an electronic format that can be E-mailed to the vendor for quote, forwarded to purchasing for acquisition, or stored on a company's Intranet for reference by maintenance technicians.

Some specialists prefer not to develop a Specification Form, but use instead a manufacturer's name and model number. This would seem to do two things: It would cut out all of the suppliers' expertise, and thereby introduce more risk into the project; second, it would give an advantage to the named manufacturer and perhaps cut out other possible suppliers. If you ask for an ABC Model 241B, that is what the vendors will supply, and you may miss advancements in their product line that may be advantageous to your application. Also, the risk that the device is correct for the application rests with you. If you ask for a device that will bend widgets, the vendor's expertise goes into his offering, and perhaps XYZ Company has a better, or perhaps, even a less expensive widget bender.

The P&ID for our project (see Figure 2-21, page 45) identifies all instrumentation and control components by symbols and tag numbers. The next step for the instrumentation and controls group is to prepare a Specification Form for each tag marked item. For demonstration, LT-100, the level transmitter in level loop LIC-100 (see Figure 2-24, page 50) will be documented.

As a starting point, we can gather some information from the P&ID. LT-100 is an electronic displacement level transmitter, to measure the level of 01-D-001, a 6' diameter tank. ISA-20 contains a Specification Form for level instruments (displacer or float); ISA form S20.26. Please refer to Figure 4-4. The instructions for the form are included as Figure 4-5 and 4-6.

## Figure 4-4: Level Instrument – Specification Form

| | | | LEVEL INSTRUMENTS (DISPLACER OR FLOAT) | | | | SHEET | OF | |
|---|---|---|---|---|---|---|---|---|---|
| | | | NO | BY | DATE | REVISION | SPEC. NO. 321 | REV. 0 | |
| | | | 0 | FAM | 12/15/2003 | | CONTRACT 1234 | DATE | 1/3/2003 |
| | | | | | | | REQ. - P.O. J-6    J-12 | | |
| | | | | | | | BY FAM | CHK'D CHK CAM | APPR. LF |

| | | # | | | | |
|---|---|---|---|---|---|---|
| | | 1 | Tag Number | LT-100 | | |
| | | 2 | Service | K.O. DRUM | | |
| | | 3 | Line Number / Vessel Number | 01-D-001 | | |
| BODY/CAGE | | 4 | Body or Cage Material | C.S. | | |
| | | | Rating | 300 psi | | |
| | | 5 | Conn Size & Location Upper | 1 1/2" TOP | | |
| | | | Type | 300 psi FLG | | |
| | | 6 | Conn Size & Location Lower | 1 1/2" BTM | | |
| | | | Type | 300 psi FLG | | |
| | | 7 | Case Mounting | SIDE | | |
| | | | Type | | | |
| | | 8 | Rotatable Head | NOT REQ | | |
| | | 9 | | | | |
| | | 10 | Orientation | LEFT HAND | | |
| | | 11 | Cooling Extension | NOT REQ | | |
| | | 12 | | | | |
| DISPLACER OR FLOAT | | 13 | Dimensions | 48" | | |
| | | 14 | Insertion Depth | | | |
| | | 15 | Displacer Extension | | | |
| | | 16 | Disp. or Float Material | 304 S.S. | | |
| | | 17 | Displacer Spring/Tube Mtl. | MFG. STD. | | |
| | | 18 | | | | |
| | | 19 | | | | |
| XMTR/CONT. | | 20 | Function | TRANSMITTER | | |
| | | 21 | Output | 4-20 mAdc | | |
| | | 22 | Control Modes | | | |
| | | 23 | Differential | | | |
| | | 24 | Output Action: Level Rise | INCREASE | | |
| | | 25 | Mounting | INTEGRAL | | |
| | | 26 | Enclosure Class | NEMA 8 | | |
| | | 27 | Elec. Power or Air Supply | 24Vdc from shared | | |
| | | 28 | | display | | |
| SERVICE | | 29 | Upper Liquid | WET GAS | | |
| | | 30 | Lower Liquid | DEGASSED MTL. | | |
| | | 31 | Sp. Gr.: Upper | Sp. Gr.: Lower | | .9 @ 60 F | |
| | | 32 | Press. Max. | Normal | 50 PSI | 4 PSI | |
| | | 33 | Temp. Max. | Normal | 400 F | 90-150 F | |
| | | 34 | | | | |
| | | 35 | | | | |
| OPTIONS | | 36 | Airset | Supply Gage | | |
| | | 37 | Gage Glass Connections | | | |
| | | 38 | Gage Glass Model No. | | | |
| | | 39 | Contact: No. | Contact: Form | | |
| | | 40 | Contact Rating | | | |
| | | 41 | Action of Contacts | | | |
| | | 42 | | | | |
| | | 43 | | | | |
| | | 44 | | | | |
| | | 45 | | | | |
| | | 46 | Manufacturer | LATER | | |
| | | 47 | Model Number | LATER | | |
| | | 48 | | | | |

NOTES:

© 1981 ISA    ISA FORM S20.26

**Figure 4-5: Level Instrument – Instructions, Part 1**

# 16 Level instruments (displacer or float)

Instructions for ISA Form S20.26.

1) Tag No. or other identification.

2) Process service.

3) Line number or vessel number on which cage or body is installed.

4) Material of chamber and/or mounting flange.

5) For float specify top or side of vessel connection. For displacer in a chamber specify upper, then lower connection; such as side-side, side-bottom, top-bottom, etc. Give flange size and rating or NPT size.

6) Same as 5.

7) Refers to position of case when viewing the front of the case relative to the chamber; the case is either to the left, right, or top.

8) On displacer instruments specify if case is to be rotatable with respect to the chamber. This only applies if there is one or more side connections.

10) Orientation of control with respect to displacer cage.

11) Cooling Extension.

13) Specify float diameter or displacer length. The displacer length is also the range.

14) Insertion depth applied to ball floats. It is the mounting flange to the center of the ball.

15) The displacer extension is measured from the face of the mounting flange to the top of the displacer. This dimension is required only for top of vessel mounted instruments.

16) Includes rod.

17) Refer to MFR's standard materials or special materials.

20) Transmitter, controller, switch, etc.

21) Air pressure or electrical signal output of transmitter or controller.

22) P: Proportional

Pn: Narrow band proportional

PI: Proportional plus Integral (Reset).

23) Differential if controller on/off must specify differential adj. or fixed. State adjustable range or fixed amount.

24) INCREASE (Direct action) or DECREASE (Reverse Action).

25) Remote, or integral.

26) Electrical classification of housing. NEMA number.

27) Air pressure or voltage. If electronic, state whether ac or dc.

---

**Figure 4-6: Level Instrument – Instructions, Part 2**

29) Used only for interface application.

30) Used for all services.

31) Specific gravities at operating temperature.

32) Operating and max. pressure, or vacuum.

33) For cryogenic service, give minimum temperature.

36) Airset assumed mounted to case.

37) Connections on chamber, give size.

38) Specify gauge glass, if required.

39) Contact form: SPST, SPDT, etc.

40) Give Volts, Amps.

41) Describe contact action with level.

47) Model number of entire assembly.

---

Starting at the top of the form, we fill it out as follows:

1. Tag number – LT-100

2. Service – Knock Out Drum

3. Line No./Vessel No. – 01-D-001

**Body/ Cage**

4. Body or cage material – Carbon Steel
   Rating – ANSI 300 (information from the piping group)

5. Connection Size & Location-Upper – 1 1/2" top (coordinate size with piping group)
   Type – ANSI 300 raised face flange

6. Connection Size & Location-Lower – 1 1/2" bottom (coordinate size with piping group)
   Type – ANSI 300 raised face flange

7. Case Mounting – Side

8. Rotatable Head – Not required

9. This space is blank on the form and also in the instructions, so it can be used as you wish or left blank

10. Orientation – Left Hand

Orientation is coordinated between the vessel or piping group and the instrument specialist. Typically, the specialist will use the instrument supplier's orientation charts to ensure the piping group understands the clearances and orientations offered.

11. Cooling Extension – Not required

## Displacer or Float

12. Same as 9, above

13. Dimensions – 48 inch

Note: We will measure only the liquid in the center 48 inches of the tank diameter.

14. Insertion depth – Does not apply

15. Displacer extension – Does not apply

16. Displacer or float material – 304 stainless steel

17. Displacer spring/tube material – MFG STD

## XMTR/Controller

20. Function – Transmitter

21. Output – 4 to 20 mA dc

22. Control modes – Does not apply

23. Differential – Does not apply

24. Output action level rise – Increase

25. Mounting – Integral

26. Enclosure class – NEMA 8

27. Electric power or air supply – 24v dc from shared display

## Service

29. Upper liquid – Wet gas

30. Lower liquid – Degassed Material

31. Sp.Gr. – Lower .9 @ 60°F.

32. Press max – Normal – 50 psi – 4 psi

33. Temp max – Normal – 400°F. – 90°-150°F.

**Options** 36 – 45. Not required

46. Manufacturer – Later, after device is purchased

47. Model Number – Later, after device is purchased

Then, fill out the upper right portion of the form, the revision number, sheet number, specification number, and contract number. The form is now complete.

Most projects have or develop overall specifications. These specifications might be specific for instrumentation and control, or they may be much broader and include all parts of the project. They may be a few pages, or they may be a complete series of specifications for all items supplied for the project. They may be printed and bound into many volumes. You should refer to the master specifications to ensure the devices you specify are in compliance.

### Classified Areas

Many projects involve handling flammable materials and all use electricity in one form or another. Flammable material, when combined with electrical equipment, becomes a potential hazard. Figure 4-7's third bullet introduces the way this potential hazard is reduced.

---

**Figure 4-7: Hazardous (Classified) Locations**

• Chemical and petrochemical plants and petroleum refineries which use materials that are flammable

• Many control system devices use electricity

• To prevent fires or explosions, equipment use and installation must follow safe practices that are set down in codes

• Information on these codes follows...

---

The installation of all electrical and electronic equipment must follow rigid rules set forth in the National Electric Code (NEC). The NEC is the law of the land in the United States. Similar, but different codes exist in Canada, Europe and elsewhere. In the U.S., local building or electrical inspectors enforce the law. It is important that those designing a project understand how the inspector enforces the code. For example, sometimes the local inspector will not inspect small changes made to an operating facility, but will rigidly enforce rules for major modifications or for a new facility.

The extent of the potential hazard is mandated by the code in a classification system consisting of three parts: Class, Group, and Division. (See Figure 4-8.)

---

**Figure 4-8: Area Classification – Electrical**

• For plants where the presence of flammable materials are present, the National Electrical Code (NEC) has set up a classification system consisting of three parts:

**CLASS**

**GROUP**

**DIVISION**

– For example, Class I, Group D, Division 1

---

The **Class** designation denotes the generic nature of flammable materials:

**Class I locations** – where flammable gases or vapors may be present in the air in quantities sufficient to produce an explosive or ignitable mixture (e.g., chemical and petrochemical facilities).

**Class II locations** – where combustible dusts may be present in sufficient quantities to cause hazard (e.g., flour mills and coal pulverizing facilities).

**Class III locations** – where the hazardous material consists of easily ignitable flyings or fibers that are not normally in suspension in the air in quantities to produce ignitable mixtures (e.g., sawmills, textile mills).

The **Group** designation defines the hazardous material, for example:

**Group A** – acetylene

**Group B** – butadiene, hydrogen

**Group C** – carbon monoxide, hydrogen sulfide

**Group D** – ammonia, most petroleum products

**Group E** – metal dusts

**Group F** – carbon black

**Group G** – flour

**Note:** This list is only an example, the complete list is much more comprehensive. NEC also requires temperature-rating code, known as "T" number.

The **Division** designation specifies the frequency of the hazard, for example:

**Division 1**

- Hazardous concentrations normally exist or often exist
- Frequent repair or maintenance causes hazardous concentrations to exist

**Division 2**

- Hazardous only in abnormal situations
- Hazardous materials are handled, but normally confined
- Nonhazardous areas because of forced ventilation
- Dust layers may accumulate

A **Division 1** location is one where:

- Hazardous concentrations exist continuously, intermittently or periodically under normal operating conditions
- Hazardous concentrations exist frequently because of repair or maintenance operations or leakage of equipment
- Breakdown of equipment or process failure might simultaneously release hazardous concentrations of flammable gases, vapors, or dust, and cause failure of electrical equipment

A **Division 2** location is presumed to be hazardous only in abnormal situations, such as the result of an accident when process equipment or a container fails. A Division 2 location is one in which:

- Flammable liquids or gases normally confined in closed containers or systems are handled or processed

- Areas normally not hazardous because of forced ventilation become hazardous if ventilation equipment failed

- Areas adjacent to Division 1 areas, where hazardous concentrations could be communicated unless prevented by positive ventilation with adequate safeguards, or

- Areas where layers of hazardous dust can accumulate

An area classification plan is developed, usually by the project electrical group, which classifies by class, group and division all areas of a plant that handles flammable materials. The wiring in an area must follow NEC practices set forth for that area's classification. Of particular importance in this chapter is all equipment installed in the hazardous areas must be labeled as suitable for that area. The device must carry the label, usually imprinted on the manufacturer's permanent identification label.

The area classification is furnished to the vendor by a nationally recognized laboratory under criteria furnished by the labeling authority. The largest and most accepted labeling authorities in the U.S. are Underwriters Laboratory (UL) and Factory Mutual (FM). The Canadian Standards Association (CSA), is also frequently accepted as a regulatory authority. The CE mark from Europe is not normally accepted by itself in the U.S., but many instruments carry all four labels. The label will list the classifications for which the device is rated, and that must match the classification of the area in which it will be installed – for example, Class I, Group D, Div. 1. Also available in the vendor's literature for the device, and possibly on the label, is the National Electrical Manufacturers Association (NEMA) type as defined in NEMA Standard 250-2003. The NEMA type applies to the enclosure of the device. Figure 4-9 describes the NEMA type of enclosures for hazardous locations.

Many instrumentation and control field devices require very little power to operate. For this reason, they can be safely installed in hazardous areas by making use of the Intrinsic Safety rating.

### Figure 4-9: NEMA Standard 250-2003

**NEMA enclosure types for hazardous locations**

| Type | Indoor/Outdoor | Suitable For |
|------|----------------|--------------|
| 7 | Indoor | Class I, Div. 1, Gr A,B,C,D |
| 8 | Both | Class I, Div. 1, Gr A,B,C,D |
| 9 | Indoor | Class II, Div. 1, Gr E, F, G |
| 10 | – | Mine Safety and Health Administration, 30CFR, Part 18 |

- See NEMA 250-2003 Enclosures For Electrical Equipment (1,000 Volts Maximum) for complete information.

---

**Figure 4-10: Intrinsic Safety**

- **Basic premise**
  - It is possible to construct a circuit that is incapable of storing and releasing enough energy to cause ignition of a hazardous atmosphere under normal, abnormal, or fault conditions
- **Implementation**
  - The energy supply to the intrinsically safe circuit is limited by a barrier
  - Energy storing components are prevented from storing too much energy
  - Or, the devices in the circuit are proven to be inherently, intrinsically safe and are employed in a properly designed intrinsically safe system

---

Figure 4-10 describes the concept of intrinsic safety. The basic premise of intrinsic safety is that, by limiting the power to field devices to a level that cannot cause ignition of a hazardous atmosphere, the field devices may be installed in general purpose housings, simplifying the overall installation. The power to the field devices is limited by specially designed wiring systems and installing barriers designed specifically for this use.

---

1. ISA-20-1981, Specification Forms for Process Measurement and Control Instruments, Primary Elements, and Control Valves (Research Triangle Park, NC: ISA – The Instrumentation, Systems, and Automation Society)

## CHAPTER FIVE

# Purchasing

## Overview

Let us assume the project design phase is finished and the stack of specification forms, or data sheets, is complete. It's now time to acquire the instruments and controls. A typical acquisition process might consist of the seven steps shown in Figure 5-1:

Many instrumentation and control (I&C) practitioners will not need to perform all the listed steps. Some of us are able to draw the instrument or valve from the storeroom without having to consider how it came to be there. Others will have been involved in the purchasing process and have some experience selecting acceptable suppliers, soliciting bids, evaluating proposals, selecting a successful supplier, and requisitioning and purchasing components.

| Figure 5-1: Seven Steps of Acquisition |
| --- |
| 1. Develop an acceptable suppliers list |
| 2. Assemble the bid packages |
| 3. Send the bid packages out for bid |
| 4. Formally receive the proposals |
| 5. Evaluate the proposals |
| 6. Award the purchase order |
| 7. Receive the devices |

Refer to the list in Figure 5-1, Seven Steps of Acquisition, to which steps you have made. At your place of employment, there may be no need for the lengthy evaluation process. However, some day you may work on a project where all this has to be formally resolved – hence, the following discussion.

## Seven Steps of Acquisition

Figure 5-1, Seven Steps of Acquisition, is an outline of the work necessary to get a control component to your facility. The process involves at least three separate entities, as well as the instrument supplier: (1) an I&C design group, (2) a purchasing group, and (3) owner representatives. All of these entities may have definite opinions whom the suppliers should be.

I&C design group has experience from previous projects to develop a list of acceptable suppliers. They will have an understanding of advances in your industry and in the latest capabilities of instruments and controls. The design group should have some understanding of unresolved problems that particular manufacturers may have experienced in the recent past. In your organization, the I&C design group may not see priced bids during the evaluation process. The feeling is the technical evaluator should focus only on the technical features of the proposal. This approach may be less common since we are all involved in getting the best value for the money expended. However, it is generally true the I&C design group wants the highest possible quality and the most features for their I&C devices.

The project purchasing group looks at the business experience of each supplier. Does this supplier have the capability to deliver the devices needed in the time available, manufacture the quality required, and provide the necessary technical and administrative support? The purchasing group wants acceptable quality at the best possible price.

Owner representatives may have had both good and bad experiences with some of the potential suppliers. If our project is an expansion of an existing facility, the owner may have definite ideas whom the suppliers might be.

Considerable diplomatic skill and a cooperative spirit among the three groups is often necessary to proceed through the Seven Steps of Acquisition in a timely manner.

### Step 1: Develop an acceptable suppliers list

On our project, we are going to competitively bid supplying a group of I&C devices. We are assuming no supplier pre-selection has been made. Preparing an acceptable list of suppliers is a bit more difficult than it probably sounds. Diplomacy helps! Three or more suppliers are needed on the "acceptable suppliers" list for truly competitive pricing.

A minimum number of bid packages is desirable, since preparing and handling bid packages takes time and costs money. As an I&C professional, you understand the market in your area and in your industry. You know the instrument supplier representatives in your area. You know which I&C suppliers have experience in your industry. Suppliers really shouldn't be on the list if they don't have the knowledge and experience to handle the special challenges of your industry or facility.

Your facility might have special requirements. For example, your process may call for devices to be cleaned for oxygen service; you many need to have sanitary connections on your instruments; or your industry may have standardized on special surface treatments and cleaning requirements for ultra-pure water systems, or possibly special diaphragm seals for pulp stock applications.

What are the criteria for being an acceptable supplier? If it is an existing plant, the owner's opinions are very valuable. Today, the differences in the capabilities and performance of I&C devices furnished by different suppliers might actually be negligible (although the manufacturers may not agree with that statement).

So, what sets suppliers apart? Investigation and thought are required if the owner has never seen a supplier at the existing facility, or if the supplier is unreasonably remote from the new facility. Personal service is an important feature not every supplier can, or will, furnish. If you have no history with a particular supplier, but a good history with other suppliers, you might want to think twice about including the unknown company on the acceptable suppliers list.

In some industries and areas of the country, it is not uncommon for instrument suppliers to check in with maintenance technicians weekly, or for manufacturers' representatives to run impromptu classes in applications, software configurations, or to make presentations on feature advances. If you have an installed base at your facility, the suppliers should check in from time to time to see how their devices are performing. The importance of local support is obvious; include it as a part of your bid evaluation!

The project's standard cover letter, or request for quotation (RFQ) letter, may include a statement that reserves the right to award the purchase order to anyone – meaning the low bidder will not necessarily receive the award. This provides significant latitude in selecting the successful supplier. However, all bidders must have an even chance of success. If you are "only going through the steps" knowing full well you will award the order to "Brand X", you are doing a disservice to the other bidders. If the decision to go to a particular supplier has already been made, it is more efficient and honest to work with your purchasing department to negotiate with the supplier of choice, without going through the bid process. It is a terrific waste of time, effort and good will to require bids from suppliers while knowing they will never be successful. That being said, if you are willing to change suppliers at your facility, your company can occasionally reap some savings during competitive bidding.

Many, if not most, facilities standardize on a single instrument manufacturer for a particular type of I&C device within a facility. This simplifies purchase, installation, training and stocking of spares. Clearly, when you have one family of magnetic flow meters, your maintenance group will be familiar with mounting the device. The face-to-face dimensions are the same, and the wires all connect to the same component in the same place. It will be easier to train the staff in configuring and troubleshooting a single instrument. Standardizing on a single supplier may allow your facility's management to enter into special, reduced pricing agreements with the supplier.

## Step 2: Assemble the bid packages

A typical list of bid packages for a large project might include the following groupings: (The list is for illustration only.)

- Pressure Transmitters, including Pressure, Differential Pressure
- D/P Flow, and Flange-Mounted Level
- Magnetic Flow Transmitters
- Vortex Meters
- Turbine Flow Transmitters

- Level Transmitters - Ultrasonic
- Temperature Elements
- On-Off Control Valves
- Modulating Control Valves - Globe and Ball
- Modulating Control Valves - Butterfly
- Control Panels or Distributed Control Systems

For effective bidding, individual packages should only group those devices normally furnished by the listed bidders. That is, don't package pressure transmitters with mag meters in one bid package unless all the bidders on the list normally furnish both devices. On the other hand, if all your proposed suppliers have, in their product line, magnetic flow meters, pressure transmitters and, for arguments sake, vortex flow meters, then go ahead and make one package for all the devices. Remember that some suppliers will not submit a bid if they can't furnish the complete package, so be sure of their scope and capabilities.

### The Bid Package

The bid package is the group of documents sent to potential suppliers to guide them in preparing a proposal to furnish a product. In its simplest form, the bid package might be a faxed Specification Form. In its most complex form, a bid package might be 27 volumes of design documents and specifications used to design and build a billion-dollar refinery. For purchasing instrumentation, a bid package probably contains a cover letter explaining what you want from the supplier, called a request for quotation (RFQ). This letter will contain:

- Contact names, address and phone number for the company that will purchase the devices

- The name and address for the entity that pays the invoice – the "Bill To" address

- Delivery location, called the "Ship To" address

- Schedule: states when the devices are needed

- Bid data: Bid due date, what documents should be included with the bid, estimated duration of the evaluation process. Documents needed with a bid should be listed and may include requests for warranty statements, installation, operation and maintenance manuals, technical bulletins and dimensional drawings. Note that "Certified" drawings needed by the piping designers are not normally furnished until after the purchase order is issued.

Finally, with all the organizational information out of the way, the bid package should, of course, include the ubiquitous Specification Forms.

In addition to the cover letter, a bid package includes the following components:
1. Invitation to Bid

The Invitation to Bid may be part of the cover letter, or possibly a separate document furnished by project management or by the purchasing team to cover this standard information:

   a. Who is requesting the quotation? The supplier should know whom to contact for questions. They will also want to establish which company will purchase the devices, "whose paper" will be used. The supplier may do a credit check on the purchaser. Unfavorable credit history may have an impact on pricing.

b. What is the project name? Suppliers will want to know if the quote is for ongoing work at the facility or if the quote is for a "grass roots" project. Sometimes, suppliers have access to special pricing for project work, so this could work to your advantage.

c. Where will the devices be used? The ultimate destination or, simply, the installation point, should be defined. Some suppliers portion out credit and profit for a sale within their organization to the representative supporting the design effort and to the representative supporting construction and operation. It is helpful to define the players up front.

d. Dates – When is the bid due – also known as the "Closing Date"? When do you expect to evaluate the bids? When will you award the order? When do you expect to take delivery? And, when do you expect to install the product? For larger bid requests, the supplier should be given from two weeks to a month to prepare the quotation. Be sure to include mail delivery times for sending and receiving the bid package. Simple quotes can be prepared in a day or so, and today, more and more quotations are delivered using E-mail. The various dates all have an impact on the suppliers' proposal workloads and on their production schedules. If the suppliers need to block out a production window to meet your requirements, they need to know that up front.

e. The letter should indicate what you want back from the bidder. At the very least, price and delivery are needed, but bid packages may include a quotation sheet which requests a statement of compliance from the bidder on specification items and commercial requirements. Occasionally there is a request to provide an un-priced quotation and a priced one, used when the technical bid evaluation is separate from the commercial evaluation. You may want specification bulletins, installation, operation and maintenance manuals, or dimensional drawings to be included with the bid to facilitate your evaluating the proposals. Sometimes, orifice plate calculations and valve calculations performed by the device supplier can be included with the bid. All your expectations should be called out in the RFQ.

2. Specification Forms

a. Include a table of contents, so the bidder knows a complete set of forms was furnished for bid. It is not unheard of for the photocopy machine to miss copying a Specification Form.

b. Check that legible copies of the Specification Forms are included, preferably using the ISA format described in Chapter 4. A completed ISA Specification Form will provide sufficient information to bid the component. Erroneous information is occasionally included which conflicts with what the Specification Form requests. For example, we have even seen construction specifications added to instrument bid requests, which prompted calls from the bidders asking if they were to install the instruments as well!

Specification Forms may be prepared in at least the following three ways:

1. **Generic.** The Specification Forms are sufficiently complete so vendors may offer firm quotations. Generic forms are used when competitive bidding is used to select or to pre-select instrument vendors. The pre-selection might take place before all technical details are known.

2. **Vendor-specific.** Vendor-specific Specification Forms are based on a specific manufacturer. They include, quite probably, the manufacturer's name and the model number of the device. Vendor specific forms are used when the vendor has been selected.

3. **Partially complete.** Specification Forms are prepared with all technical data and project requirements included, but without the manufacturer's name and the device model number. Potential vendors complete the form and the information supplied by the successful bidder is used to complete the Specification Form. Partially complete Specification Forms may be used to select a vendor or may be used after a vendor has been selected.

Generic data sheets are prepared when there is concern that listing a device manufacturer gives one supplier an advantage over another. More than one potential supplier has declined to bid when seeing their competitor's name, or even features pointing to that competitor, listed on the specification data sheet. It takes time, and therefore money, to prepare a bid. If the potential supplier feels he or she has a low probability for success, that supplier may very well decline to expend the effort.

Competitive bidding on instruments may be falling out of favor anyway. Many corporations have partnered with their instrument and controls suppliers, so purchasing new devices may simply mean calling the supplier on the phone and invoking a corporate buying agreement. Or, for facilities that have a history with particular manufacturers, the purchasing department may just negotiate a pricing structure based upon historical data and goodwill.

Interestingly, there may be a hidden advantage to generic data sheets. The design professional has more opportunity to concentrate on the process information, materials and required performance of the instrument instead of expending design effort figuring out the manufacturer's model number codes and options. The more effort expended determining the minimum, normal and maximum conditions the device will operate under, the fewer the surprises during operation and the better the system. Operating ranges are important!

## Tagging

Do you buy your instrument identification tag as part of the device model code? Many suppliers include that option. Many specification forms include a standard note that calls for the device supplier to "Permanently affix a stainless steel (SS) tag engraved with the device tag number given above", or some similar phrase. The device supplier dutifully checks a box on their factory order sheet and somewhere down the line a tag is added to the device or one is tossed into the shipping container. Consequently, the facility ends up with as many different styles of instrument tags as there are suppliers of devices.

Sometimes, the facility ends up with MORE tag styles than instrument manufacturers – the factory tag is different than the tag furnished by the local supply house! Occasionally, but assuredly, instruments will evade the process altogether and will arrive on site without any tag or with an incorrect tag. The furnished tag will be tossed out with the shipping box, or it will magically fall off the correct device and get reaffixed to the wrong device later. There must be a better way.

Remember, in most cases the separate tag that was requested on the Specification Form is an option from the device supplier, so there is an associated cost. We say "separate" tag because many instruments and control valves will already include a nameplate riveted to the device, engraved by the manufacturer with the manufacturer's logo, the model number, the serial number, and our owner-furnished tag number. These nameplates may not be particularly easy to read in the field after some months of operation, but they are perfectly legible for warehousing, calibration and construction.

The cost for the optional separate tag can be significant – $25 and up in some cases. For your money, you end up with all types of tags, as well. In our experience, these have ranged from an engraved phenolic plastic plate to a piece of thin SS tape hand scratched with the tag number. Furnishing your site tag specification to an instrument vendor is rarely successful or inexpensive. They are in the business of manufacturing high quality instruments, not tags.

It makes a lot more sense for the end user to specify and purchase all instrument tags at one time, to one specification, and from one supplier. This way the end user can control the material, engraving method, font size and attachment method. The tags are sure to meet the specifications because one entity furnished them. We have seen successful systems where the commissioning or startup team affixed tags as part of their procedures – the presence of a tag flagged a device as ready for operation.

The cost of an SS engraved plate with heavy-duty SS wire can be a fraction of the tagging option cost from the instrument supplier. Another successful tagging option uses plastic plates with a printed adhesive bar code in addition to the bold tag printing. The bar code quickly links the physical device to the site database for review, or specification data and calibration history, and it simplifies maintenance record keeping.

## Step 3: Send the bid packages out for bid.

Bundle up the bid packages and send them out, or forward them to the purchasing group which will send them out. Many companies severely limit contact with potential suppliers during the selection process, so there is no possibility of unfair advantage given to one bidder over another. It sounds easy, and it is, to prepare a bid package. The following are some suggestions for a smooth bidding process based on past "opportunities", or, in other words, really bad experiences.

1. Verify the bidder's name and address. The business card you used for the address might be several years old. Or, the salesperson that calls on you spends all his or her time on the road and rarely picks up their mail. The bid package may need to go somewhere other than the mail drop address on the card.

2. Request the purchasing group to notify the recipient that the package is coming. Then ask them to wait a few days and call again to confirm its receipt. More than one bid was never received by a potential supplier. Mail sent is not necessarily mail received.

### Step 4: Formally receive the proposals

The professional way to handle proposals is for one group, often the purchasing group, to hold all the bids until the date and time mentioned on the Invitation to Bid. At the stated time, the packages can be opened and distributed for evaluation according to your company's rules. The evaluation is often an activity for the three groups – purchasing, I&C, and client representatives. The purchasing group frequently performs the bid opening with witnesses. The goal is to give no bidder an advantage over another, to offer no pricing hints, or to give any bidders grounds to distrust your impartiality. Contents and pricing of the bids are confidential and should never be shared with the other bidders. The bidders may ask you, as is natural, "How did I do?". It would be unfair to the other bidders to respond. Forward all such calls to the purchasing professionals.

### Step 5: Evaluate the proposals

There are two components to a bid evaluation – the commercial and the technical. The I&C group will do the technical evaluation. The purchasing group will do the commercial evaluation. They will address the warranty offered by the seller, the FOB call-out (the transfer of ownership during shipping), and other points included in the "Terms and Conditions".

The more likely involvement of the I&C professional is for the technical evaluation. Here, the proposed instrument is compared against the specification. Does it meet the requirements set forth on the Specification Form? A spreadsheet is a great tool for making this comparison. On the left column, the specification items are listed from the Specification Form: the performance, metallurgy, connection size and method, accuracy and repeatability, range, and anything else of importance from the Specification Form.

In each subsequent column, the specific offering from each bidder is listed, with the evaluators' determination if the offering meets the specification. When the evaluation is complete, the columns are complete. It is not uncommon for all of the bids to meet the specification requirements. In this case, the technical evaluation would state that the bids are technically equal. Then the selection will be made on purely commercial terms.

Care must be taken to evaluate against the Specification Form – not against some attractive, but unnecessary feature offered by one bidder that exceeds the requirements. Necessary features are specification items, unnecessary features are, well, expensive fluff.

## Shipping

The cost of shipping, or freight, for your instrument can be a significant, if somewhat hidden charge. While you as the person using the device may not care how it got into your hands, you probably should care if the company that signs your check is spending money that could otherwise go to more important stuff like your pay or health plan. It is not uncommon for freight to be 15% of the cost of an instrument, or more. Understanding these costs will allow you to make better decisions on how quickly your instruments are delivered and how much it costs to get them into your hands.

Generally, the faster the shipment, the more expensive it will be. Ground transportation is less expensive than air freight. Also, the bigger the order, the lower the "per unit" shipment cost. It takes almost the same amount of paper-work to ship a pallet of instruments as it does to ship one instrument. On a large shipment the cost of preparing the shipping documents, the invoice, and the accounting is spread across many items instead of one. With a little fore-thought and, to be realistic, with a fair amount of luck, sending large orders early enough to allow ground transport should realize some shipment savings. Furthermore, special handling charges can be significant. Suppliers will understandably want to share that extra cost with you if they have to disturb their normal production runs to manufacture your special widget.

The bid provided by your friendly neighborhood instrument supplier should have mentioned how freight will be handled. Some of the common freight payment terms are listed below.

| PPA | Prepay and Add – The supplier pays the shipping, but the cost is then added to your invoice so the purchaser (you) ultimately pays for freight. This is probably the most common freight payment method. |
|---|---|
| Allowed | Allowed – The supplier pays for shipping. |
| Collect | Collect – The shipping company will collect the costs upon delivery. |
| Own Truck | Own Truck – There are several variations on the term, but it means that the instrument supplier will deliver it to you. There is usually no cost, unless a cost is stated. |
| Will Call | Will Call – You will pick up the instrument at the supplier's location. Many suppliers have a door that is actually marked "Will Call". Will Call is, of course, more common for smaller components that can be hand carried, and when your local supplier is truly local. |
| Owner's Truck | Owner's Truck – again, there are many variations on this term, but it means you will pick up the device. Similar to Will Call, but used more for larger components that require a truck and loading dock. |

Instrument suppliers can be understandably reluctant to use Freight Allowed terms, since they can't predict if expensive overnight shipping will be called for later – after they have committed to a price. Shipping Collect is more expensive than other pre-paid methods. Third party shipping companies need to be paid for their time and effort expended to collect money at the receiving end, and for the possibility that no one will pay for the shipping after the fact. PPA freight is more common. The shipper pays freight costs at the time of shipment and gets the best shipping rates. The suppliers are happy because they don't have to predict the unpredictable shipping costs.

### Step 6: Award the purchase order

The purchasing group probably will make the award based upon results of the technical and commercial evaluations. Some observations about the process:

1. It is always a good idea to review the purchase order against the Specification Forms or your bid evaluation. It is not unheard of for items to get forgotten, for substitutions to be made, or other problems to arise. A cursory glance at the documents may allow you to catch problems early, while they are still small.

### Payment Terms

Instrument and controls engineers and maintenance personnel normally have little or no involvement with the legal contents of a purchase order. However, for the curious, the following explanation discusses the process and some of the common terms used.

The "terms and conditions" is part of a purchase order. Indeed the "T&Cs" are frequently pre-printed on the purchase order form itself. It establishes the legal agreement between your company and the company that supplies the device. You, as the maintenance or design person, may be asked to comment upon the duration of the device warranty, or you may be involved with setting up the required delivery schedule. But other than those "component usage"-based aspects of the agreement between the two companies, there may be little need for I&C involvement.

However, if you are in a hurry for your instruments, you will want to encourage your purchasing group to ensure the T&Cs, or any other legal aspects of the purchasing process, are in order. It is appropriate to take great care when the devices you want to purchase are critical to your schedule – or if you are new to the bidding and purchase of I&C devices.

Disagreements between the buyer and seller are not uncommon, and they can add weeks to the acquisition process. Preparing a formal "Request for Quotation" package that includes final specification forms and a copy of your companies "standard" terms and conditions is good way to surface problem areas early in the bidding process, leaving time for the purchasing and legal departments to resolve any conflicts.

A word of warning when working with a design or construction contractor: don't assume your company's purchasing relationship with the supplier will apply when the devices are purchased by another entity – even if it is for your facility. The company selling the devices is only concerned with whose name is on the purchase order, not with the destination. It may not matter if your company has a long history with that supplier, the legal agreement between the seller and the buyer, in this case the contractor, will still need to be worked out. Credit checks will be run, prior history will be reviewed. The discount pricing that your company enjoys will have to be negotiated if the devices are bought under someone else's purchase order. Sometimes the contractor may even have better pricing for a specific project than your company receives. We have even been in the situation where the contractor had to work with the owner to have the owner clear an outstanding debt before the instrument supplier would ship the instruments. Pricing negotiation takes time, so the earlier you address the arrangements, the more time there will be to rectify any problems.

The payment obligation term appears as some variation of the words "Net 30". Net 30 means the buyer will pay for the device within 30 days of invoice. "Invoice" is the bill for the device, which is probably sent to your company the same day the device is shipped. The "Terms and Conditions" may key off other dates – for instance, the payment duration may commence the day the device ships from the factory or it may start the day you receive the device, regardless of the day the invoice arrives. Some payment schedules, particularly for more expensive or fabricated items, may call for staged payments based on some fixed duration after receipt of the order (ARO). For example, the supplier might look for 15% of the order value in payment two weeks after receipt of the order (to cover design costs), 30% four weeks after order (possibly to cover raw material costs), etc.

2. The delivery schedule is based upon the shipment duration provided in the quotation. However, the bidder probably included the letters "ARO" when the delivery duration was listed. ARO means "After Receipt of Order". Sometimes that means "Written Order" so the lag in mail delivery may become important. Other times, a phone call is sufficient to start the order. These opportunities for misunderstanding between the buyer and the seller should all be worked out with the supplier prior to the purchase order award. In most cases, having a purchase order number is sufficient to start the order.

3. The quoted shipping duration was probably based on the "Ex-Works" or a "Ship" date, NOT a "Received By" date. Yet, many times the schedule assumes the Ship date and Received By dates are the same. When your instruments are traveling across the county, same day shipping and receiving is pretty unlikely with current technology.

## Step 7: Receive the devices

Your facility or project probably has some sort of receiving system in place to keep track of delivered components. If the project is part of an existing facility there may be conflicts between the project group and the existing facility receiving group. The project group may want all materials received and documented separately at a segregated facility. The existing facility receiving group may want all materials received through their normal channels at their normal location.

To prevent conflict, detailed procedures and arrangements for receipt and disbursement should be defined and agreed upon in advance of any material deliveries. There will be a list of purchase orders and a way to keep track of individual items on the order. I&C devices are easy to track, since they are uniquely identified with a tag number. The receiving process involves noting the purchase order and the tag number of all devices received, then getting that information to "accounts payable". The project accounting group will pay the invoice when the components are received, in accordance with the purchase order terms and conditions. As I&C people, we just need to make sure the lists and logs are accurate. Make sure appropriate entries are made when instruments are removed from the storage area. Dates are important also since the invoice process starts with shipment or delivery of the device.

Instruments are frequently shipped with the operation manuals in the shipping container. Care should be taken to ensure the manuals are delivered to the correct party. In our experience, copies of manuals should go primarily to the facility maintenance group, followed with delivery of any extra sets to the construction personnel. The construction personnel need Installation Details and Loop Diagrams for mounting the devices. It is the plant startup personnel and the owner maintenance personnel who need the detailed information in the manual. Understand too, the purchase order may have included a limited number of manuals; replacement manuals can be expensive. The trend now is towards free access to manuals that are downloaded using the Internet.

In the previous pages, the seven steps of selection have been discussed in some length. Problems can easily occur if the I&C group, the purchasing group, and the client representative do not agree to agree during the total I&C device selection process.

Your authors know of instances where entire requisitions were never processed and where items somehow mysteriously disappeared from purchase orders. Perhaps the epitome of non-cooperation was a project where the entire set of I&C devices was ordered, not once but twice. Of course, one set was surplus upon receipt at the job site. Unfortunately, there was also some personnel "surplus" after this occurrence. Don't let it happen to you.

## CHAPTER SIX

# Logic Diagrams

## Overview

Most continuous process control schemes include "on-off" control. The on-off control may involve the action of a simple switch or it may entail a long series of steps comprising a complex automatic system. On-off control can start and stop a single motor or it can be used to initiate an orderly shut down of an entire plant upon detection of an unsafe condition.

A simple system might back up an analog control system to ensure a tank does not overflow. A complex system might scan for many different process conditions and equipment states to protect major equipment worth billions of dollars from damage. Some on-off systems are software based and others are hardwired. Be the system simple or complex, it needs to be documented. The proposed on-off control scheme will need to be documented and presented for review, discussion and implementation. The proposed control interrelationships will need to be recorded so programming personnel understand the intended operation.

P&IDs were developed to show a continuous process, which they do very well. However, a different presentation method is needed for on-off control.

There are at least three methods used to document on-off control. Each method can be used by itself, or, more frequently, they will be used together in some form at different stages in a design project to describe how the on-off control operates. The three methods are:

- Text Descriptions
- Ladder Diagrams
- Logic Diagrams

The success each method has depicting the operating logic is dependent on the complexity of the on-off control needed, the ability of the logic diagram designer using the symbols correctly, and the ability of the process owner, maintenance technician, and even the programmer to read the document.

## Text Descriptions

The interaction of the equipment and the action of final elements to the sensed process conditions can be described with short statements that very briefly describe the action and reaction. This set of logic statements, often called interlock notes, will sometimes appear on the P&ID, perhaps along the right side of the drawing or in a section along the bottom. A number and perhaps a symbol identify the separate statements. The symbol, typically a diamond, and the number may also appear on the P&ID next to each of the elements affected: the process sensing switches, transmitters, the valves and the motors.

This is an effective method to use because of its clarity. The method facilitates drawing reviews and, more importantly, it is a good system to jog memories years later. The operating description is carried along with the P&ID in perpetuity. Drawbacks to the method are due to the necessary brevity of the statements. There is not a lot of space available on an already busy P&ID. The P&ID statements need to be short.

Complex logic may call for a more detailed description than that done on a P&ID. Furthermore, technicians doing the control computer program normally will need more detailed information than a few notes on a drawing. For this reason, the text-based system can also occur on a separate document, sometimes called a Functional Specification or Operation Description. There are as many names for these as there are companies that use and produce them. The text-based logic documents can be prepared for each P&ID, or it may be useful to document the operation of an entire system in one document. A typical Functional Specification for an entire system will include several sections:

1. System Description – The system title and a paragraph describing what the system does, where it starts and where it ends.

2. Referenced P&IDs – The drawing numbers and titles.

3. Motor List – A listing of motors and interlocks that impact their operation.

4. Loop List – A listing of analog loops affected by on-off control elements.

There are many different applications of the text-based logic description. The document should be written by an experienced control professional. Programmers will use the document as the basis of design. Statement brevity is good; remember logic is on-off, so the statements should reflect that.

## Ladder Diagrams

A Ladder Diagram is a stylized document based on wiring diagrams prepared for relay logic circuits. Ladder Diagrams can be used to program programmable logic controllers (PLCs), the on-off control for some distributed control systems (DCSs), or the logic for hardwired relays. Ladder Diagrams show two vertical bars, the line and the neutral, connected by a series of horizontal "rungs" containing logic components: coils and contacts. The vertical lines connected by horizontal logic circuit lines have a ladder-like appearance, hence the name.

Documenting your operating logic system by printing out the PLC program is simple and accurate. The drawback is not everyone working at a process facility is able to read a Logic Diagram quickly. If programming notes were not used, reading the ladder can be very frustrating. On a PLC ladder diagram, the individual action elements, the coils, can be referenced by many individual rungs, so reading the printout can call for thumbing through hundreds of pages of printout. This can be confusing, even when they are indexed properly.

## Logic Diagrams

The Logic Diagram is the most versatile method of depicting on-off control. A little training and experience is necessary to read a logic diagram. The symbols are simple enough to understand, yet very complex operations can be shown efficiently. All the on-off control elements for a single piece of equipment can be shown clearly on one drawing. Figure 6-1 contains a definition of a Logic Diagram.

Logic Diagrams use a series of symbols to indicate what is happening in an on-off system. While there are many variations of Logic Diagrams, there is much similarity among them. The basis for the information we will use is ISA-5.2-1976-(R1992), Binary Logic Diagrams for Process Operations[1]

### Figure 6-1: Logic Diagram

• The Logic Diagram is a conceptual document that defines the on-off state of a process, and depicts the scheme necessary for control.

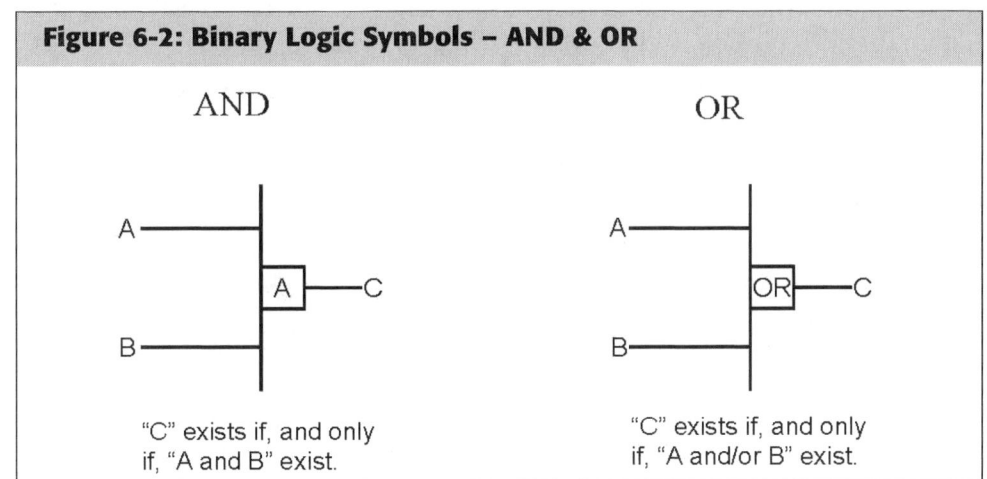

**Figure 6-2: Binary Logic Symbols – AND & OR**

AND

OR

"C" exists if, and only if, "A and B" exist.

"C" exists if, and only if, "A and/or B" exist.

Figure 6-2 shows two of the symbols used in Logic Diagrams, the "AND" and the "OR". A Logic Diagram is set up with the inputs or actions on the left side of the drawing and the result, or results, on the right side. The "AND" symbol signifies all inputs must exist (or all actions must be taken) before the result occurs. A more formal definition from Figure 6-2 is the result "C" exists if and only if "A" and "B" exist or if action "A" and action "B" have both been taken. If there are more actions feeding into an "AND", all actions must have taken place to get the desired result. There is no limit to the number of actions feeding into an "AND".

The "OR" symbol signifies that one or more inputs must exist (or one or more actions must be taken) if the result is to happen. A more formal definition from Figure 6-2 is the result "C" exists if, and only if, "A" and/or "B" exist. If there are actions "A" through "Z" feeding into an "OR", one or more of the actions must have taken place to get the desired result.

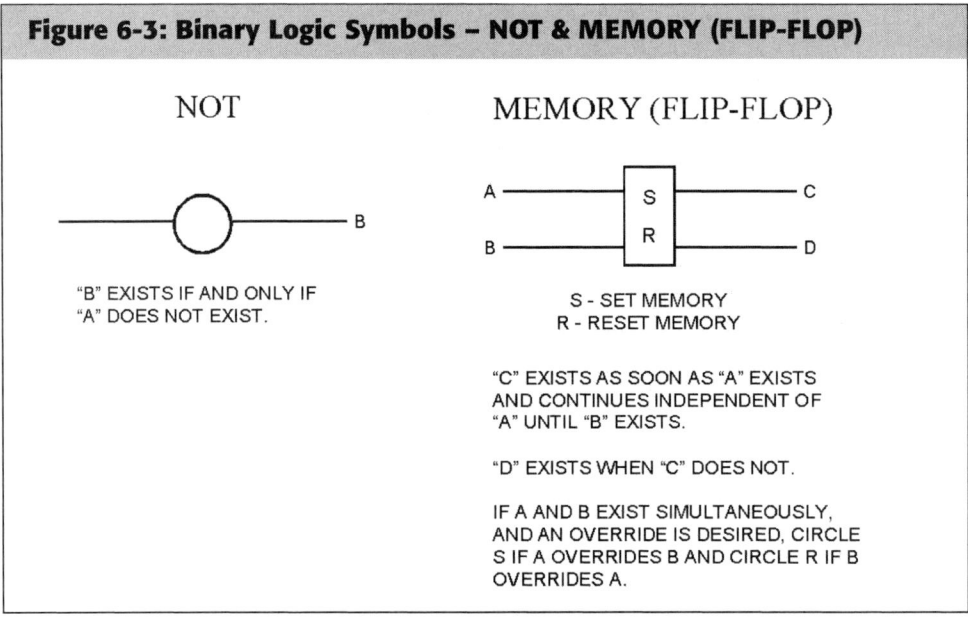

Figure 6-3 shows the "NOT" and the "MEMORY (FLIP-FLOP)" symbols. The "NOT" symbol reverses the input. If the action has taken place and is fed through a "NOT", no result will show. If the action has not taken place and is fed through a "NOT", the result will show. A formal definition from Figure 6-3 is "B" exists if and only if "A" does not exist.

The "MEMORY" symbol is more complex. If an action has been taken, the result will occur and continue to occur until another action takes place. The symbol has two outputs, and they flip-flop. If one shows the action, the other will show no result.

The formal definition from Figure 6-3 is "C" exists as soon as "A" exists and continues independent of "A" until "B" exists. "D" exists when "C" does not. If "A" and "B" exist simultaneously and an override is desired, circle "S" if "A" overrides "B", and circle "R" if "B"' overrides "A".

We have shown only a few symbols. ISA-5.2 includes many more.[2]

## Logic Devices

Several devices that can perform on-off control have been mentioned, including relay logic, PLCs and even DCSs. The oldest form of on-off control, relay logic, was accomplished by wiring electro-mechanical relays into control circuits. Relays were great for the purpose, but they can take up a fair bit of room, contacts and coils can fail, they generate a lot of heat, and they can make complex on-off control schemes difficult to implement and revise.

With the advent of electronic data processing came computers to perform the same functions as relays. The most common and versatile of these is the PLC. These have software-based "relays" and "coils" that implement our Logic Diagrams through programming. They are relatively inexpensive and reliable. As the cost of computational power decreases, more DCSs and bus-based control systems implement on-off control as well. Some of these systems use ladder diagram programming, while others use function-block-based programming which we have not discussed. Suffice it to say, they are programmed using a set of symbols that, once understood, can be read and interpreted relatively easily. Electronic controllers, such as the single loop digital controller (SLDC), have some integral on-off control capabilities. Some field transmitters even have on-off control capability. The on-off control capabilities of electronic configurable devices are almost limitless.

## Documenting On-Off Control

When do you document your on-off control? The American Petroleum Institute (API) Recommended Practice 750 states: "The mechanical design information should include…a description of shutdown and interlock systems…".[3]

The DOE Architectural Engineering Standards state: "Instruments and controls shall be documented by logic drawings interconnect drawings…ISA Standard 5.2 will be the basis of logic drawings."[4] These standards do not define logic drawing interconnect drawings.

(Note: The authors assume a comma was omitted between " logic drawings" and "interconnect drawings" in the reference, above.)

OSHA, in some new ANSI standards, specifies that safety-oriented systems be properly documented. They include shutdown and interlock systems which are designed to mitigate personnel and equipment losses .

During design, the logic documentation is a coordination tool between the different design disciplines. The equipment group, the process engineers, the operators, and  maybe the chemical engineers all have input on how the system is to be operated. For purposes of this discussion, let us call them all "process engineers".

Ideally, their input should be recorded in a manner easy for them to use and understand. Frequently, the process engineers work with the instrument and controls (I&C) group to set the entire control scheme – the analog and on-off control – using the P&IDs and one of the three forms of logic documentation. Later, the programmers work with the I&C group to implement the logic, sometimes long after the process group defined the intended operation. When the electrical group handles on-off control, the Logic Diagram can be the interface enabling the electrical group and the I&C group to complete the design of the interlock system cooperatively.

### An Example Logic Diagram

Figure 6-4 illustrates how a Logic Diagram is used to define an on-off system.

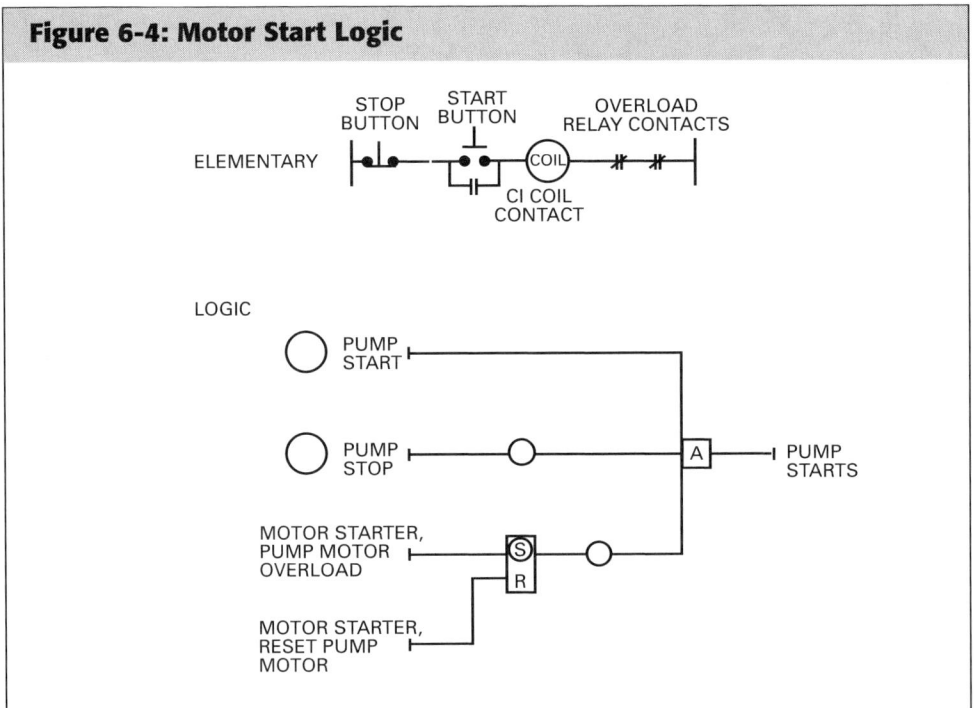

**Figure 6-4: Motor Start Logic**

This figure includes a common motor start circuit, as it would appear on an electrical elementary drawing. To start the motor, depress the start button located next to the motor or on the starter housing itself. This action completes the circuit to the coil in the motor contactor, if the overload relay contacts are closed – that is, if the motor is not overloaded. As the coil is energized, coil contact "C" closes and the motor starts. To stop the motor, the stop button is depressed. This interrupts the circuit to the contactor coil, which opens the contacts and removes power to the motor. Figure 6-4 also shows how this same action is shown on a Logic Diagram.

For the pump to start, all three inputs to the "AND" are necessary. The first action is to actuate the start button. The next action is not to actuate the stop button. This action is reversed by the "NOT", and so the top two inputs are satisfied. Note this allows you to have a separate action to stop the motor! The motor starter is not overloaded; there is no input; and, therefore, no output. However, there is a "NOT" in the line which reverses the no output and the pump starts.

In the "MEMORY", the "S" is circled. This means the starter overload overrides the starter reset. In real terms, this means if the starter shows an overload, the reset button will not start or jog the motor.

Next, we will develop a Logic Diagram for our project. Please refer to Figure 2-21 in Chapter 2, P&IDs (page 45). There are three diamonds with an internal L1 on the P&ID. One is on the output of LSL-301, the low-level switch on 01-D-001. The second is on the start-stop scheme (HS/HL-401 & 402) of 01-G-005, the condensate pump. The third is on HV-400, the hand operated on-off valve. The diamond is on the output of ZSH-400, the position switch high which signifies that HV-400 is fully open.

From the P&ID we can see there is some sort of safety instrumented system (SIS) involving a low level in 01-D-001, the condensate pump 01-G-005, and the on-off valve HV-400.

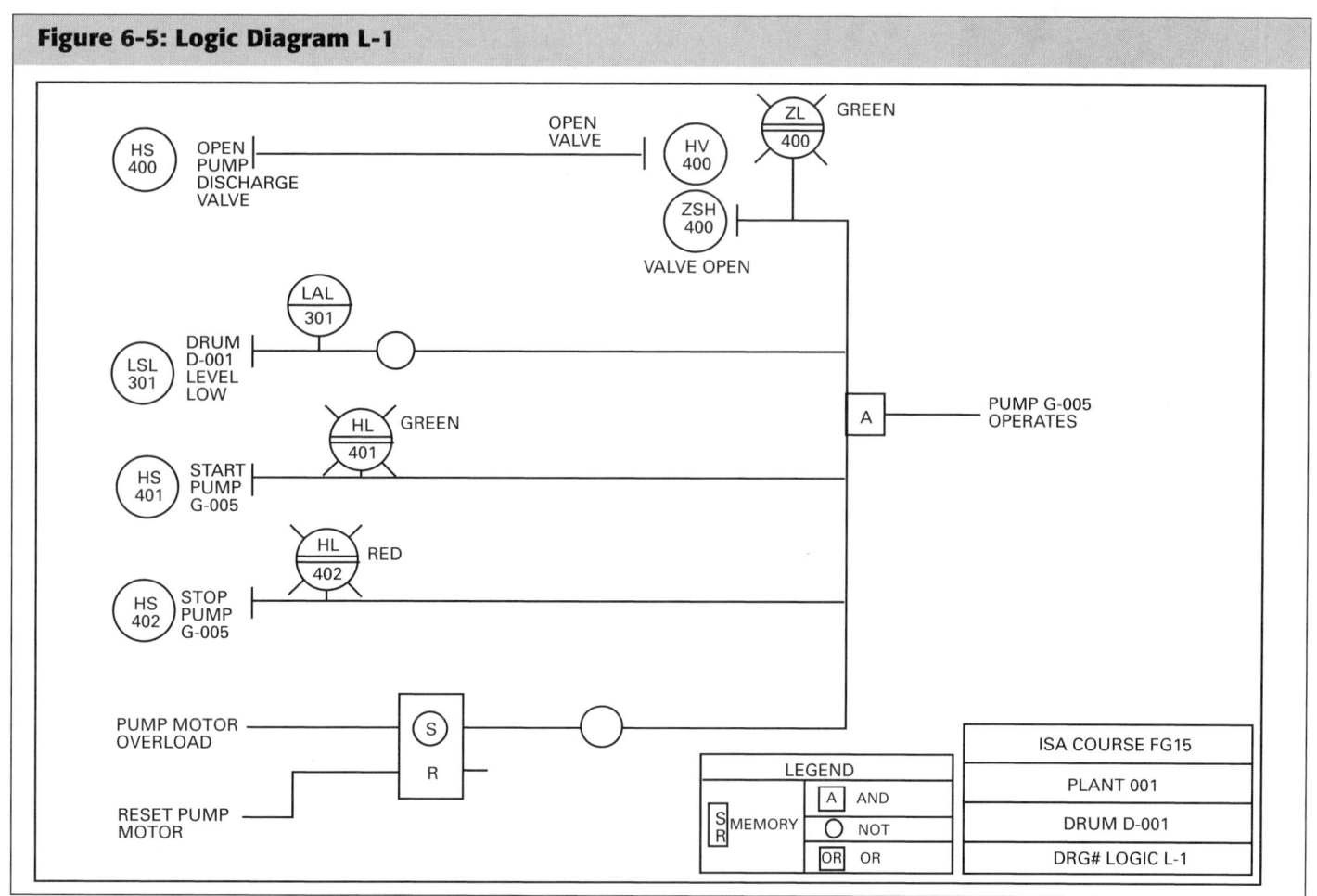

**Figure 6-5: Logic Diagram L-1**

The details are shown in Figure 6-5, Logic Diagram L-1. L-1 has five inputs to the "AND". All five must be satisfied before the pump will operate. Starting at the top, we depress HS-400 to open HV-400. There is a short time lag before the valve is proved open by ZSH-400 and ZL-400 is lit.

The next input, LSL-301, and LAL-301 is not actuated, that is, there is liquid in the tank. This action is reversed by the "NOT" and the second input is satisfied.

Next, pump start switch HS-401 is depressed, green light HL-401 is lit, and this input is satisfied.

The fourth input HS-402, the pump stop switch, is not depressed. This action is reversed by the "NOT" and the fourth input is satisfied.

The last input to "MEMORY" is not activated, so there is no output from the "MEMORY". This action is reversed by the "NOT", so this input is satisfied and the pump starts.

## Safety Instrumented Systems

Logic Diagram L-1 shows how a Safety Instrumented System (SIS) works. For a simple SIS, a word description might tell the same story. In order for the pump G-005 to start, the start button must be depressed, there must be liquid in D-001, the stop button must not be depressed, and the pump motor must not be overloaded. However, when the SIS is complicated, a word description can easily lead to misunderstandings and an unsafe plant. Therefore, in the opinion of many – but certainly not all – instrument control specialists, the use of Logic Diagrams will make plants safer. The use of Logic Diagrams will require training a critical mass of people. Members of the design teams and the owner's engineering and operations personnel must be trained to understand and use Logic Diagrams.

## SIS Can Have Many Names

So what do you call your SIS? The on-off control scheme is called by many names. Some of these names include:

- Discrete Systems
- Emergency Shutdown Systems (ESD) or (ESS)
- Interlocks
- Interlock Systems
- Protective Logic Systems
- Safety Instrumented Systems (SIS)
- Safety Interlock Systems
- Safety Shutdown Systems (SSD)

- Safety Systems
- Shutdown Systems

For simplicity, we will use the abbreviation "SIS" to describe all the on-off control of continuous process plants. Many I&C specialists will agree there are two categories of on-off control. However, the ISA dictionary definitions of the above terms do not confirm this clearly.

First, there is on-off control to run the normal operation of the plant. There are motors to start and stop, equipment to run. Failure of these control schemes is certainly undesirable; a failed on-off control scheme might make a troublesome spill. A failure might produce off-specification product. All of these results are annoying and expensive, but no one is hurt or at risk. No equipment is destroyed.

The second category introduces a different responsibility for on-off control systems – those that are safety related. The documentation used may appear quite the same, but their importance is much greater. The distinction between normal operating on-off control systems and safety systems can be fuzzy. One approach might be to use the regulatory documents to define which systems are truly safety systems and which are normal operating systems. There are many codes and standards that address on-off control. For instance, the National Fire Protection Association (NFPA) published standards for the operation of many different kinds of burners. The standards describe in detail how the on-off control is to work.

*The Automation, Systems and Instrumentation Dictionary* defines SIS as: "Safety Instrumented Systems (SIS). A system that is composed of sensors, logic solvers, and final control elements whose purpose is to take the process to a safe state when predetermined conditions are violated…other terms commonly used include emergency shutdown system (ESS), safety shutdown system (SSD), and safety interlock system."[5]

There are two general methods used to decide when an SIS and its content should be used – qualitative and quantitative. The qualitative method might be based on the skill and judgment of the project team. Some processes are well known, and what was included in the last plant design is often the basis for the current project. Previous experience of some team members might suggest certain parts of the process need special care. Books and technical papers are written describing certain processes and how they can be operated safely. Our project team has decided the reactor control will be safer if the basic process control system (BPCS) is separated from the SIS.

For improved safety, it is often desirable to separate the basic control scheme from the SIS. ANSI/ISA-84.01-1996, Application of Safety Instrumented Systems for the Process Industries, in annex B.1.1 states:

"Separation between BPCS (Basic Process Control System) and SIS functions reduces the probability that both control and safety functions become unavailable at the same time, or that inadvertent changes affect the safety functionality of the SIS. Therefore it is generally necessary to provide separation between the BPCS and SIS functions."[6]

For illustration, Figure 6-6 shows how an SIS might be implemented as an integral part of the BPCS.

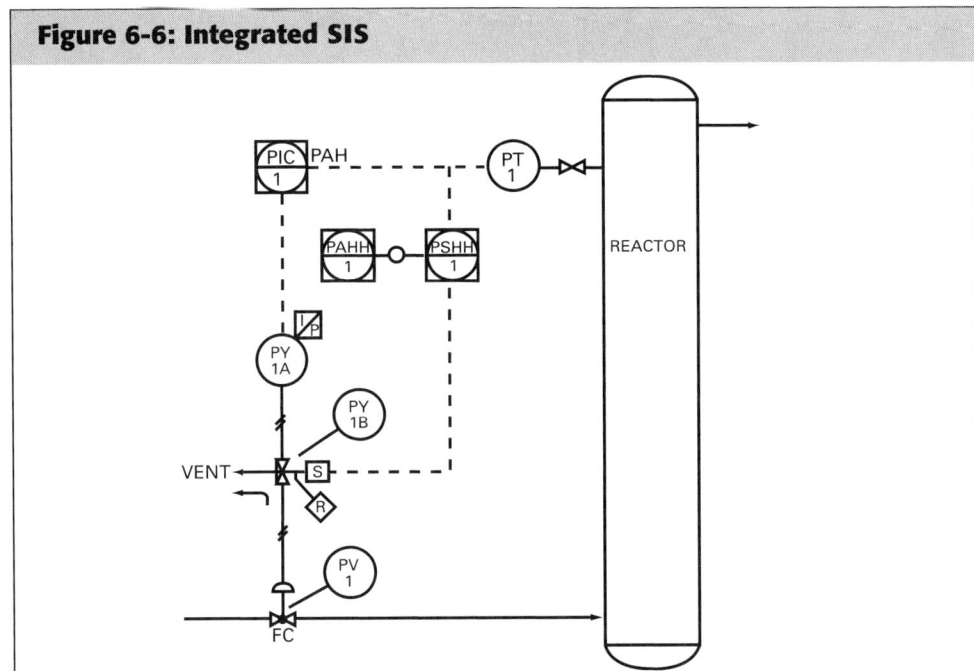

**Figure 6-6: Integrated SIS**

This figure shows a simple SIS system that employs the on-off capabilities of a shared display shared control system, or a DCS or a programmable electronic system (PES).

The SIS shuts off the feed to the reactor when the reactor pressure is too high. When the system is operating normally, PIC-1 modulates the position of the control valve PV-1, which keeps the reactor pressure at the set point. As the reactor pressure rises above the set point, a high-pressure alarm (PAH – part of the BPCS) is activated. If the reactor pressure continues to rise, a high-high pressure alarm, PAHH-1, activates. Concurrently, the power to solenoid valve PY-1B is interrupted and PY-1B moves to its fail position. PY-1B shuts off the signal to PV-1 and opens the ports to permit the pneumatic pressure on PV-1 to vent to atmosphere. Since PV-1 is a fail close valve (FC), it closes, shutting off the feed to the reactor. PY-1B stays in the failed position until the pressure in the reactor decreases below the alarm point of PSHH-1 and until the solenoid is reset locally and manually, as indicated by the "R" within the diamond.

It is apparent this system will control the reactor pressure as long as the BPCS is operational. However, a separated SIS will make the overall system safer.

Figure 6-7 shows how an extra level of safety is added to the system by separating the BPCS and the SIS. A new pressure switch, PSHH-2, is added to the scheme. This switch is connected directly to the reactor. In addition, solenoid PY-2 now controls the pneumatic pressure to a new control valve, PV-2. The control valve is fitted with a position switch ZSL-3 to confirm the valve is closed. In normal operation, loop PIC-1 controls the pressure in the reactor. An alarm sounds if the reactor pressure rises above the set point of PAH-1. PAH-1 is an integral part of the BPCS. If the reactor pressure continues to rise to the alarm point of PSH-2, valve PV-2 closes, the feed stops and the reactor pressure will rise no longer. When the BPCS is operational, solenoid PY-2 can be manually reset and the BPCS and the SIS are in service again.

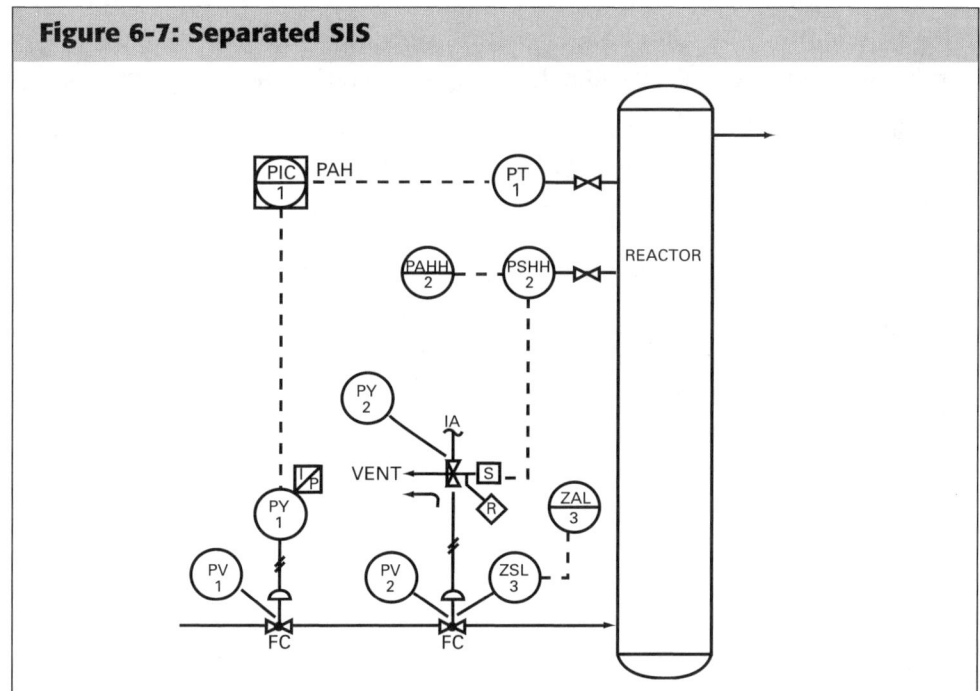

**Figure 6-7: Separated SIS**

We have used a written description of the system to explain how the reactor SIS and BPCS work.

An alternate to the qualitative method is the quantitative method. This method is described in ANSI/ISA-84.01-1996[7] and IEC 61511[8], a process industry specific standard.

The quantitative develops mathematically answers to the following questions:

- What is the risk of failure – the probability of a system failing to respond to a demand?

- What are the consequences of that failure? If the system does not respond correctly, what happens?

- Are these consequences more than are acceptable? If yes, then develop a new SIS to reduce the risk, and repeat the process. If no, proceed with the implementation of the SIS as designed.

The ISA and IEC standards help in reaching numerical answers to the above questions. The quantitative method may sound difficult to implement. However, some firms that have adopted this method are reporting excellent results. Some firms also are reporting lower costs for safety systems and safer plants. The emphasis is placed where the consequences are the most severe by the use of a mathematical solution to the analysis.

## Summary

In this chapter we have looked at on-off control in a continuous process plant and how the Logic Diagram can help in our understanding of how a SIS (Safety Instrumented System) works. We defined what on-off control is and how Text Descriptions, Ladder Diagrams and Logic Diagrams can document it. We offered examples of Logic Diagrams and pictured some of the symbols that may appear on one.

We compared the electrical elementary drawing with the Logic Diagram for motor start circuit. We described an integrated SIS for a reactor and how the same reactor might be controlled by a separated system.

Lastly we discussed briefly the quantitative and the qualitative methods of developing a SIS using ANSI/ISA 84.01-1996 as a basis.[9]

1. ISA-5.2-1976-(R1992), Binary Logic Diagrams For Process Operations (Research Triangle Park: ISA – The Instrumentation, Systems, and Automation Society, 1976[R1992])

2. ibid.

3. API Management of Process Hazards - Recommended Practice 750 (Washington, DC: American Petroleum Institute, 1990) p.3.

4. DOE-ID Architectural Engineering Standards, Instrumentation, Section 1665, Paragraph 1665.3 - Identification and Drawings, Paragraph 3.2; available from http://www.gov/publicdocuments/doe/archeng-standards; Internet; accessed July 2003.

5. *The Automation Systems and Instrumentation Dictionary*, 4th edition (Research Triangle Park, NC: ISA –The Instrumentation Systems and Automation Society, 2003) p.433

6. ANSI/ISA-84.01-1996, Application of Safety Instrumented Systems for the Process Industries (Research Triangle Park, NC:  ISA – The Instrumentation, Systems, and Automation Society 1996)

7. ibid.

8.  IEC-61511-1,2,3. Functional safety-Safety instrumented systems for the process industry sector, *Parts 1,2,3* (Geneva, Switzerland:International Electrotechnical Commission 2003)

9. op cit ANSI/ISA - 84.01

# Loop Diagrams

## Overview

The Loop Diagram is surely the most recognizable document used in the instrumentation and controls field. It is also the most common document, except for perhaps the Specification Form (instrument data sheet). The popularity of the well-executed Loop Diagram stems from its universal utility. Loop Diagrams are found on designers' desks, in construction trailers, in notebooks in the maintenance shop, stuffed in the pockets of engineers and maintenance technicians, and they can even be found littering the bottom of field junction boxes. They are spread throughout industrial facilities because people use them. Loop Diagrams tell us what components comprise our control system and how they are interconnected in a stylized but logical and somewhat elegant fashion. The presentation of information on Loop Diagrams, and, indeed the information itself, is undergoing some change and development as the systems we connect to change as well.

The ISA dictionary definition of a Loop Diagram is: "A schematic representation of a complete hydraulic, electric, magnetic or pneumatic circuit."[1] The complete circuit is generally called a loop. Two definitions of a loop, also from the ISA dictionary, are:

"A combination of one or more interconnected instruments that are arranged to measure and/or control a process variable...

"All the parts of a control system: the process, sensor(s), transmitter(s), the controller, and the final control element."[2]

These definitions are broad and cover any interconnection between two devices.

The key concept in loop definition is that all the devices in the loop monitor or control *a single process variable*. A simple loop might be a field mounted pressure switch that illuminates a light on a local panel when the pressure in a pipeline is too high or, for you purists, when the process pressure is above the switch set point. This switch may reside on its own loop or it may be included on the loop that controls the line pressure. A "pipeline" was chosen for this example since it is common for the pressure control loop to be in another operating unit, or off site. The single element pressure switch loop is familiar.

### Interconnecting a Single Instrument?

The authors, like many of you savvy readers, have had a problem visualizing what is meant by "one interconnected instrument". We suggest the "interconnected" single device assumes interconnection to some other device, perhaps a process control computer, or a device such as a now dated mag meter which consisted of two interconnected components, a flow element FE and a transmitter FT. This makes sense since, traditionally, a stand-alone or self-contained instrument like a pressure regulator or a rotameter does not call for its own Loop Diagram.

The most common electronic control loop consists of a transmitter, a controller, an I/P transducer, and a control valve. A more complex electronic system that is still a single loop includes a transmitter with two or more control valves with I/P transducers, as long as the valves exist to influence a single process variable.

Figure 7-1 and Figure 7-2 show typical loops.

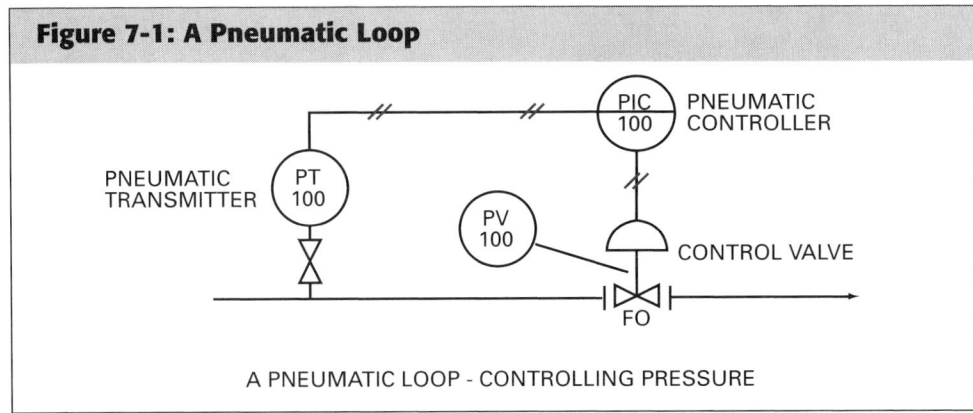

**Figure 7-1: A Pneumatic Loop**

A PNEUMATIC LOOP - CONTROLLING PRESSURE

Figure 7-1 shows a pneumatic loop (PIC 100) controlling pressure in a pipeline. All signal transmission is pneumatic at 3-15 psig.

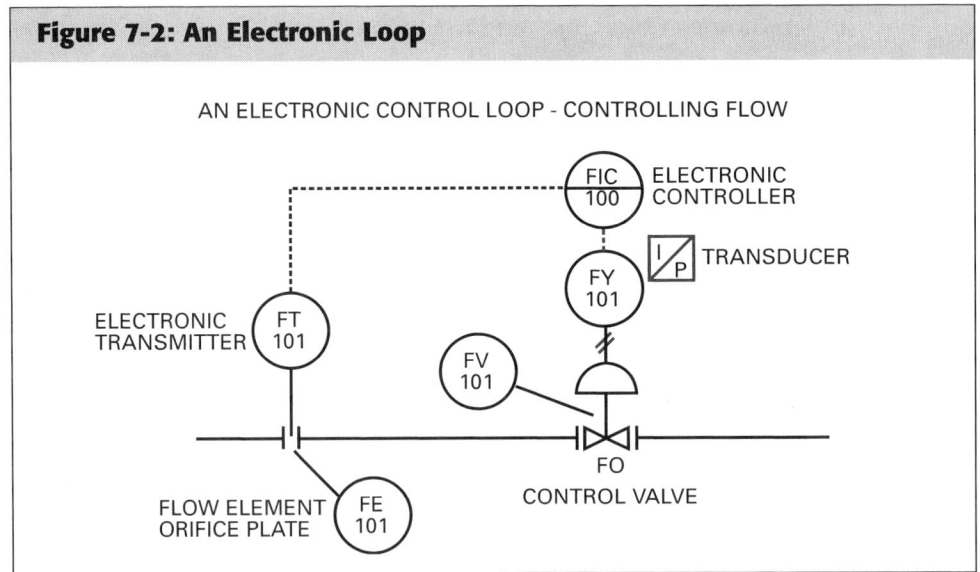

**Figure 7-2: An Electronic Loop**

AN ELECTRONIC CONTROL LOOP - CONTROLLING FLOW

Figure 7-2 is an electronic loop (FIC 101). The transmitter and controller outputs are electronic signals, most commonly 4-20 mA. The transducer (FY 101) changes the electronic signal to pneumatic, typically, 3-15 psig. Sometimes a higher-pressure signal, 6-to-30 psig, is used to provide more power to stroke the control valve.

The loops are defined on the P&ID, meaning the components that comprise a single loop are assigned during the development of the P&ID. Therefore, by the time the Loop Diagrams are developed, the components have already been numbered.

So what comprises a proper Loop Diagram? A Loop Diagram depicts the instrumentation and control devices that influence, control or indicate a single process variable. All electrical and pneumatic signal and power connections for those devices should be shown and identified along with the connections to the process – the piping, ductwork, or vessel. A Loop Diagram defines the inter-connection of each device, including the control computer or pneumatic controller. A Loop Diagram identifies the source and conditions of utilities like air, electric power and water. The Loop Diagram provides termination information including junction box identifiers, terminal strip numbers, terminal block numbers and even pneumatic port identification. Depending on the requirements of your facility, cable numbering, conductor colors, wire markers, power panel call outs and circuit breaker numbers all can be found on a Loop Diagram.

## Guidelines for Loop Diagrams

ISA-5.4-1991, Instrument Loop Diagrams, establishes a few guidelines for Loop Diagrams.

- Loop Diagrams are typically ANSI B size, 11" by 17", drawings. The B size drawing is used primarily because it has just the right amount of space to depict a single loop. This is a perfect size because it is easily handled, both in the field and on the technician's bench. As a bonus feature, this size drawing can easily be reduced to ANSI A size, 8 1/2" x 11", for desktop reference and even storage in a 3-ring binder. Many laser printers can produce 11" x 17" drawings. So it is simple to print off a copy of the Loop Diagram before you dash out to figure out why the operators lost the flow signal (probably because someone closed the upstream manual block valve, if your day goes like ours).

- Loop Diagrams depict only one loop. This guideline may sometimes appear to be violated for drawings of alarm panels and multipoint indicators where a tabular approach may be used. However, despite the loop-like appearance of these drawings, they are not, strictly speaking, Loop Diagrams, since they violate the cardinal rule of a Loop Diagram: all the devices in the loop monitor or control *a single process variable*.

- Loop Diagram format may be either horizontal or vertical, but the chosen format should remain consistent for the facility or operating unit. Horizontal formats are more common.

- Loop Diagrams are divided into sections reflecting the component locations. Sections used in the Loop Diagram examples of ISA-5.4 include:

  1. Field or Field Process Area
  2. Cable Spreading Room
  3. Computer I/O Cabinet/Cabinet
  4. Panel Front
  5. Panel Rear
  6. Control Panel
  7. Console

### Loop Sheet Size Variations – a Rant and Rave

You may be one of the unfortunates that have to live with Loop Diagrams published on a bed-sheet-sized drawing, ANSI E size, with 4 or 6 loops to a drawing. This defeats the primary benefit of the Loop Diagram – one loop per drawing. This is actually helpful to NO ONE, except possibly some long departed project design team. They had to follow an unfortunate contractual anomaly that priced drawings by size – they got paid less for six 11" x 17" drawings than they did for a 36" x 42" drawing.

Someone may have subjected your facility to this abomination because they thought it would save design dollars. What this really means is your design contractor had too much power or your management had too little understanding of process control requirements. So, you need to educate your management.

Luckily, with today's computer aided drafting, the bed sheets can be cut up into individual Loop Diagrams fairly easily, but someone is going to have to pay for it!

The specific sections used in your facility may be different from those listed, but a split between field and control panel is fairly standard. The section approach is a good one whose usefulness has been proven by long experience. Panel Front (or primary location normally accessible to the operator) and Panel Rear (or normally inaccessible to the operator) sections have become common with the advent of distributed control systems (DCSs). Local (or auxiliary location) control panels are frequently included in the field section rather than in a dedicated area. Your goal is to pack sufficient information into a small area. The use of too many section breaks can cause inefficient use of space. One should establish layout standards with care. Three major sections is probably a decent compromise.

- Loop Diagrams may show a minimal amount of information, or they may be extremely complex and packed with lots of information. Minimum and optional information sets are discussed later in this chapter. Of course, the content of the Loop Diagrams at your facility should follow a site standard. The amount of information maintained on Loop Diagrams has a real, direct and significant impact on your cost of design and operation, so there are contractual and work-hour consequences to choices made regarding Loop Diagram complexity.

Here are some additional guidelines not directly addressed in ISA-5.4, but which have proven useful to include or discuss:

- Although not shown in ISA-5.4, Loop Diagrams, a schematic representation of the process equipment is helpful in both locating the devices and to quickly understand the loop's function. There is, of course, a cost associated in showing the "process cartoon". The P&ID is a good source for the "process cartoon" shown, but it needs editing to remove extraneous information.

- The title of the Loop Diagram is normally the service description that appears in the Instrument List, on Specification Forms and elsewhere on design documents and, sometimes, on the shared display screen.

- Stating what should be obvious, the drawing number of a Loop Diagram is the loop number, since that is the most natural way people will search for the document. Unfortunately, there can be resistance to this numbering scheme, since the loop number sometimes does not fit into the drawing numbering rules used for a project or a facility. In the authors' experience, it is simply the responsibility of the instrumentation and controls design team to find a way to keep the Loop Diagram numbering system based on the loop number, as no other system makes sense.

## Optional and Additional Information

- Sometimes optional information specified in ISA-5.4 is included on the Loop Diagram. This can be particularly helpful for installing and troubleshooting, but again, there is a cost adding and maintaining this information. Furthermore, adding this information to a Loop Diagram means that information appears in more than one place, the Loop Diagram and other project documents (perhaps the Specification Form). Duplicating information is normally strongly discouraged. Having data in two places means you will probably, at some time, consider the information to be correct nowhere. However, the advantage of having manufacturer's data and calibration information on the Loop Diagram cannot be denied. Some operating units decide to maintain the Loop Diagram as the primary document for the instrumentation and controls group, so the Specification Forms become a secondary document used only for purchasing during project work. In other words, the Loop Diagram is the controlled document. Care is taken to ensure its information is always correct. The Specification Forms may not be as carefully maintained (but they certainly should be).

- Despite what you may have heard, device location information may be shown on a Loop Diagram, although this is not shown on ISA-5.4. Hey, they are your drawings; you can put whatever you want on them. While it is not normally done, there are facilities that find floor and column location callouts for field devices to be useful. While this data can be helpful for the maintenance people at 4 a.m. when trying to track down a failed instrument, one must understand location information is particularly expensive to provide and maintain. It may be available from a Location Plan (see Chapter 8). Device locations are not regularly available at the appropriate time in the design sequence. They may have to be added after the Loop Diagram is otherwise complete. The extra, out of sequence effort is expensive. Also, instruments tend to move during design, so there is a surprisingly large cost to monitor and correct locations shown on Loop Diagrams. This is probably better information for the Location Plan and database, but that is a decision to be made by the project team or facility management, with the understanding there are contractual implications.

- Additional information is included on an electronic or shared display Loop Diagram when it is drawn to the ISA-5.4 definition of optional items. Therefore, it is important to know whether the Loop Diagrams in your design package include the minimum required information or the minimum plus optional items.

This information originates from the Instrument List, Specification Forms, Location Plan, Installation Details, and other project documents. All this information and more could and should be included in a project database.

## Loop Diagram Development

Now that you have an idea of what you want to include in the Loop Diagram, it is time to make them. In general, Loop Diagram development can commence when:

- Design criteria for Loop Diagram content is agreed upon – in other words, you have a standard!
- P&IDs are issued for design.
- Specification Forms are complete.
- Devices are specified or, better yet, purchased.
- Field junction box and marshalling panel architecture is understood and defined.
- Control computer termination diagrams are ready for "loading".

## Typical Loop Diagrams

The following figures and discussions explain some typical Loop Diagrams and the information they contain. It should be noted that ISA-5.4 is the only generally accepted international standard that describes Loop Diagrams. It is not a complex document. It presents six typical Loop Diagrams and a few symbols used in their development. The six typical Loop Diagrams are of varying complexity and include two each for pneumatic control, electronic control and shared display and control. One of each type shows the minimum required items, and the other shows additional optional items. The symbols used on Loop Diagrams combine those used on P&IDs from ISA-5.1, but with additional electrical and pneumatic connection information.

The additional information is shown on the next three figures. Remember, when developing Loop Diagrams, the goal is to schematically define ALL the connection points from the field devices to the process control device, be they pneumatic or electronic. All junction boxes, marshalling panels, pneumatic cabinets and terminations should be uniquely defined. All the connection information needed to install, maintain and troubleshoot the loop components should be shown.

Figure 7-3, Loop Diagram - Terminal Symbols, shows symbols used for terminals, bulkheads and ports. This should duplicate the identification placed on the device by the manufacturer. It is important that the termination callouts used on the Loop Diagram match that provided on the instrument. Showing + and – on the Loop Diagram may be correct for the 4-20 mA signal, but it won't be helpful when the device terminals are actually called out as 4 and 5, or when there is a signal + and – as well as a test + and –. The opportunity to connect to the wrong termination may appear to be silly now, but it will happen nonetheless. Ambiguity will initiate misunderstanding. Misunderstanding

**Figure 7-3: Loop Diagram – Terminal Symbols**

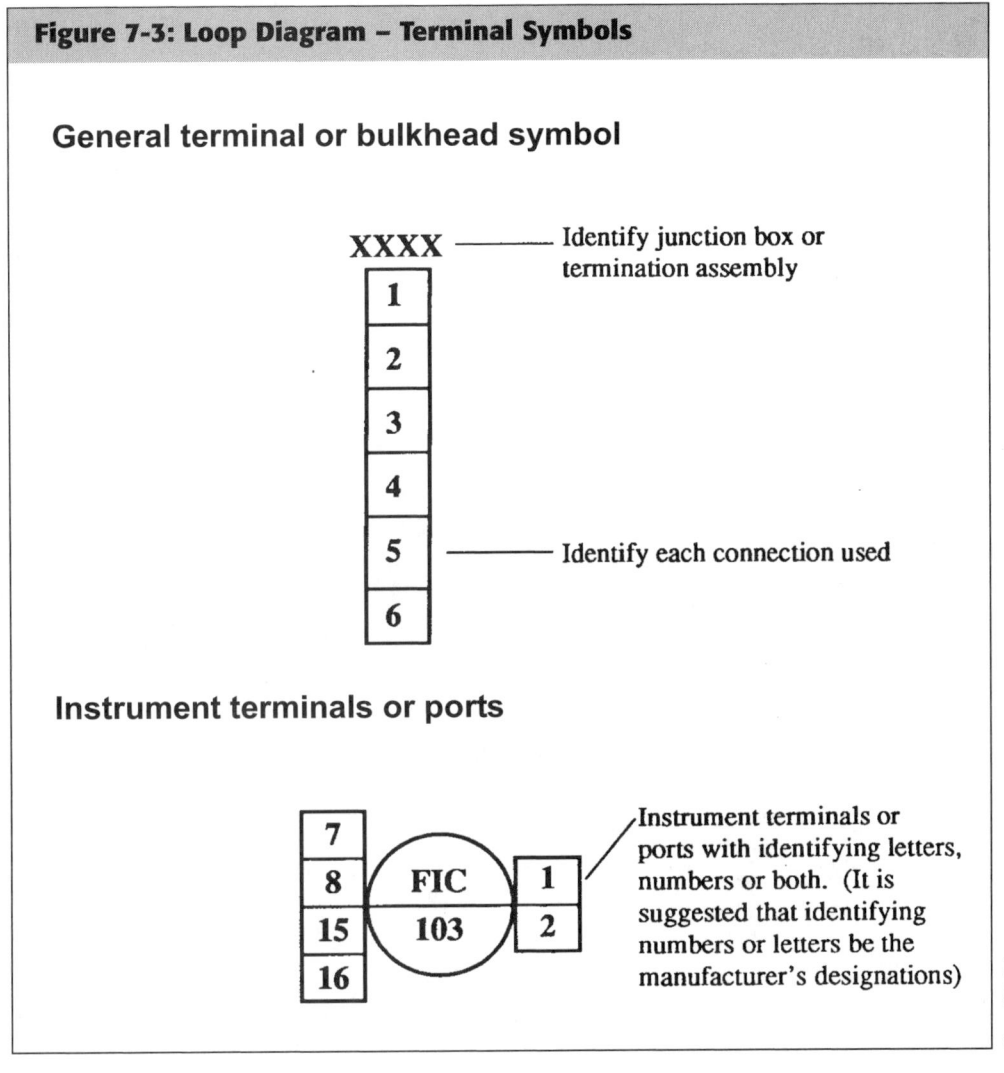

**General terminal or bulkhead symbol**

XXXX ——— Identify junction box or termination assembly

1

2

3

4

5 ——— Identify each connection used

6

**Instrument terminals or ports**

7

8   FIC   1

15   103   2

16

Instrument terminals or ports with identifying letters, numbers or both. (It is suggested that identifying numbers or letters be the manufacturer's designations)

*From ISA-5.4*

during installation induces errors. Errors during installation cause needless expenditure of VERY expensive hours, operations personnel goodwill, and maintenance technician energy during startup.

Figure 7-4, Loop Diagram - Energy Supply, shows how to connect the energy sources to the loop: the electrical source, air supply, and hydraulic fluid supply. Electrical connections show the number of conductors, voltage level, and panel and circuit numbers of the source. Air or hydraulic supply shows the number of connections and supply pressure. Since the Loop Diagram is schematic, the physical connection of the instrument is NOT shown. Physical requirements of the installation appear on the Installation Detail that defines the physical layout, tubing size and fittings, wiring components and all other materials needed to mount the device.

Figure 7-5, Loop Diagram - Instrument Action, identifies the response of the device to applied signals, be they pneumatic or electronic. A process transmitter or controller can be supplied in either of two actions, direct or reverse. In a direct acting controller or transmitter, the output signal increases as the process variable increases. In a reverse acting controller or transmitter, the output signal

**Figure 7-4: Loop Diagram – Energy Supply**

## Instrument system energy supply

**Electrical power supply.** Identify electrical power supply followed by the appropriate supply level identification and circuit number or disconnect identification.

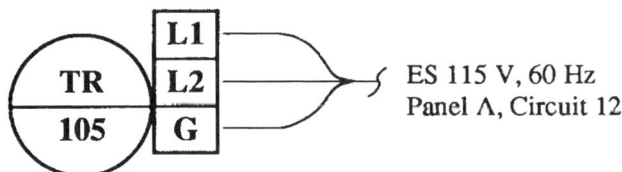

ES 115 V, 60 Hz
Panel Λ, Circuit 12

**Air supply.** Identify air supply followed by air supply pressure.

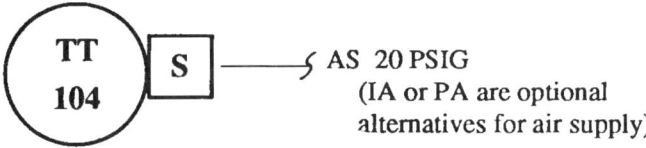

AS 20 PSIG
(IA or PA are optional
alternatives for air supply)

**Hydraulic fluid supply.** Identify hydraulic fluid followed by the fluid supply pressure.

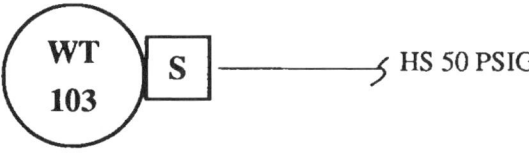

HS 50 PSIG

**Figure 7-5: Loop Diagram – Instrument Action**

**Identification of instrument action.** Show the direction of the instrument signal by placing appropriate letters close to the instrument bubble. Identify an instrument in which the value of the output signal increases or changes to its maximum value, as input (measured variable) increases by the letters "DIR". Identify an instrument in which the value of the output signal decreases or changes to its minimum value, as the value of the input (measured variable) increases by the letters "REV". However, since most transmitters are direct-acting, the designation DIR is optional for them.

decreases as the process variable increases. The action for controllers should be indicated on a Loop Diagram as DIR or REV. An incorrect controller action will make a control loop unstable. That instability may not be noticed until the loop is in operation. If the supplied action is incorrect, it is not difficult to reverse a controller action. The DIR for transmitters is not normally shown since they are almost all direct acting.

In Figure 7-1, A Pneumatic Loop, PIC 100 will be used to describe the use of a direct or reverse acting controller. Transmitter PIT 100 is specified on its Specification Form as a direct acting device. The output signal increases as the pressure in the line increases. Because control valve PV 100 fails open, an increasing output from PIC 100 will close the valve. Therefore, the controller must have reverse action, and the loop will work as follows:

As the pressure in the line rises, the output of PIT 100 rises. Acting on this rise, the controller output decreases, opening control valve PV 100 and lowering pressure in the line. Figure 7-1 duplicates how loop PIC 100 might look on a P&ID. The DIR or REV is, therefore, not shown on Figure 7-1 and would not be shown on a P&ID.

## Loop Diagram for PIC 100

For our project, we have developed a Loop Diagram - PIC 100, Figure 7-6.

Refer to Figure 2-21 P&ID (Chapter 2, page 45). PIC 100 is a pneumatic field mounted pressure controller which maintains the pressure in 01-D-001 at 4 psig by releasing excess vapors to the flare through butterfly control valve PV 100. We determine the range of 0-10 psig for transmitter PIT 100 on line 10" 150 CS 004 from the Specification Form. From the transmitter vendor print we find it has two ports, an air supply port "S" and a pneumatic output port "O". The controller PIC 100 is located at elevation 114'6" - N-1075' 0", E-1500'0". This location and elevation as well as information for PT 100 and PV 100 can be obtained from the Location Plan. See Chapter 8, Figure 8-4.

From the Specification Form we learn that PIC 100 is a reverse acting controller with three ports. The vendor print shows the three ports, an input port "I", output port "O", and supply port "S". The pneumatic output signal goes through tube number PV-100-(B) directly to the valve positioner port "I". There are two other ports on the positioner, an output "O" and a supply "S".

Note that a valve positioner does not show on the P&ID. It is only shown on the Loop Diagram. Valve positioners often do not have a unique tag number because they are mounted on, and supplied with, control valves. The three air supplies originate at the six-connection manifold located at N 1060'0", E 1500'0" as shown on the Location Plan, Chapter 8, Figure 8-4.

**Kids – Don't try this at home:**

Many times, rules are appropriate. One situation that calls for a rule of thumb, or a rule in general, is the specification of valve positioner and I/P transducer action. In a modern control system, there are plenty of opportunities to fool with device action, at the transmitter, at the controller, at the valve. It is probably a really good idea to specify that all positioners and I/P transducers are direct acting, if only to limit the degrees of freedom in your system.

You will appreciate this at 4 a.m. when trying to troubleshoot a misbehaving control loop. There is nothing worse that trying to get your head around a system with a several changes in controller action from the transmitter to the control system, through the control algorithms, to the controller output, to the I/P transducer, to the valve positioner and then onto the valve itself.

**Figure 7-6: Loop Diagram – PIC-100**

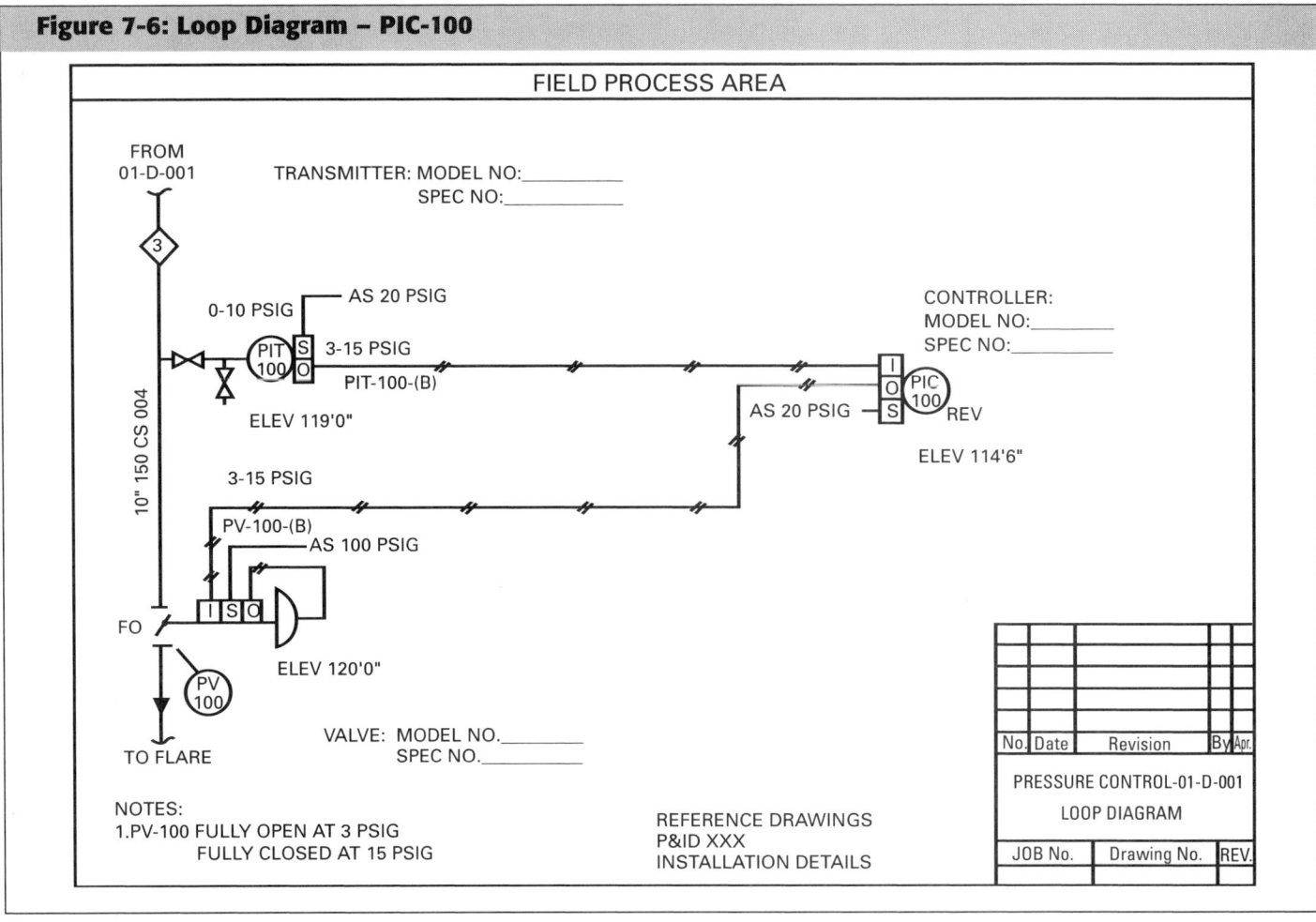

The loop works as follows: PT 100, a direct acting transmitter, sends an increasing signal as the pressure in 01-D-001 rises. The rising signal causes reverse acting controller PIC 100 to decrease its output to the direct acting positioner on butterfly control valve PV 100. Because PV 100 fails open, the decreasing signal opens the valve and lowers the pressure in 01-D-001.

## Loop Diagram for FIC 301

Figure 7-7, Loop Diagram - Electronic Minimum, shows an electronic loop FIC 301. On the left side of the drawing are the items in the field process area. The same tag numbers and symbols are used as are on the P&ID, but the symbols include additional connection details. Orifice plate FE 301, flow transmitter FT 301, and control valve FV 301 are shown. A pair of wires, FT-301-1 and FT-301-2, connect the + and − terminals of the transmitter to terminals 1 and 2 of the field junction box JB 30. There is a shield to protect the pair from electromagnetic interference. The shield is isolated at the transmitter as indicated by "shield bend back and tape" and connected to a common terminal 3 at JB 30.

A multiconductor cable 10 connects JB 30 with junction box JB 40 in the cable spreading room. The purpose of the cable spreading room is to reor-

**Figure 7-7: Loop Diagram – Electronic Minimum**

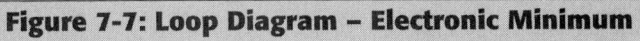

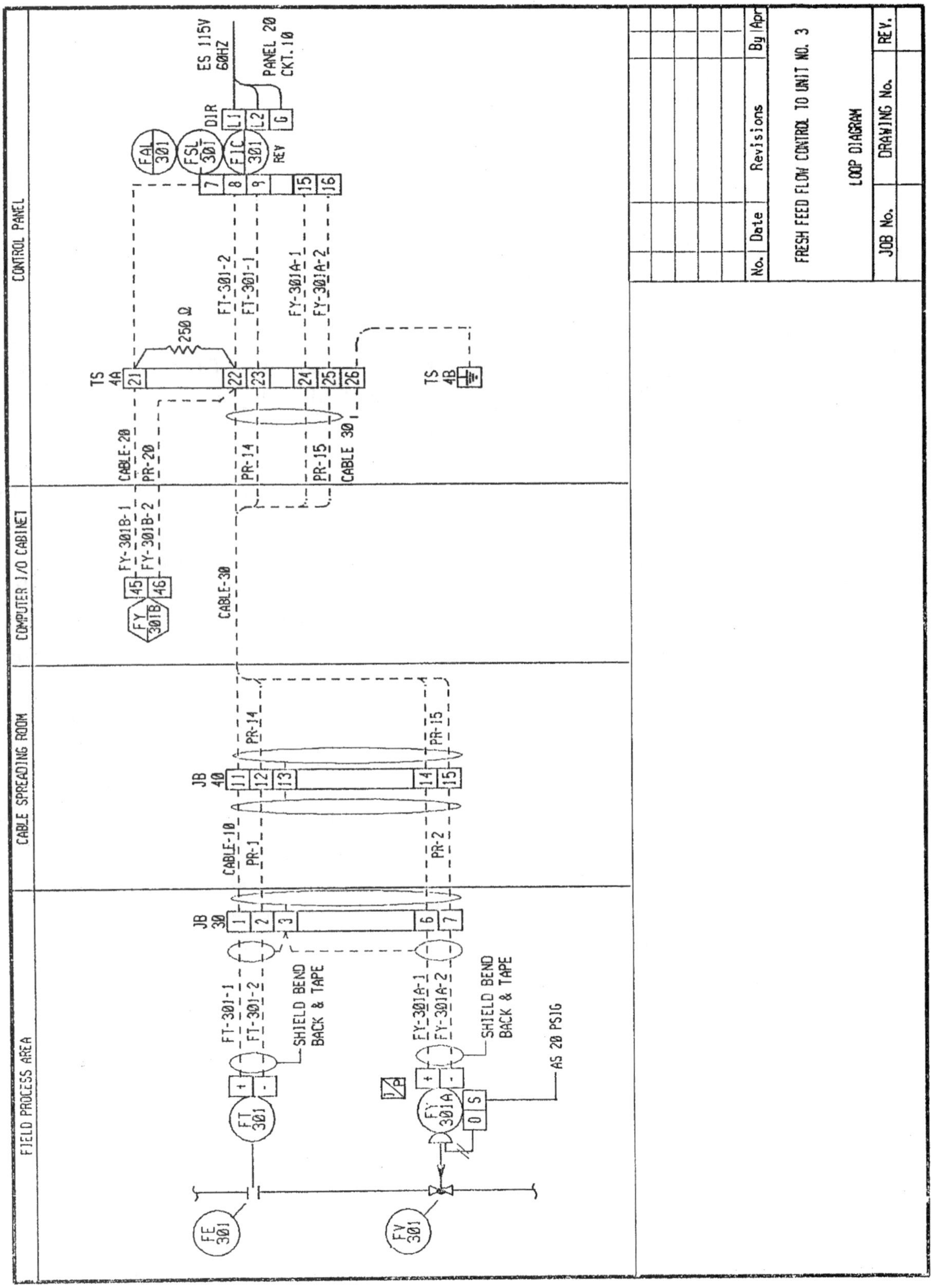

**Loop diagram, electronic control, minimum required items.**

*From ISA-5.4*

ganize the wires from the field multiconductor cable 10 to an optimal arrangement for the control panel multiconductor cable 30. The overall shields for the multiconductor cables are connected to terminal 13 of JB 40 to provide a continuous shield from the field to the control panel. The shield of cable 30 is connected to ground through terminal 26 on TS 40, and from there to the "ground" terminal on TS 40.

There is an input signal from FT 301 to the computer FY 301B. The 250 ohm resistor connected to TS 40 terminals 21 and 22 converts the 4-20 mA signal of the flow transmitter to 1 to 5 volts needed by the computer. This signal uses pair 20 of cable 20, wire numbers FY 301B - 1 and 2. FIC 301 has a "behind the panel" integral low flow switch, FSL 301, and accompanying "front of panel" alarm FAL 301. FSL 301 is a direct acting switch. As the flow signal decreases, the output decreases and the alarm sounds. FIC 301 is powered by 115 volt 60 Hz from circuit 10 of panel 20. The power supply is connected to FIC 301 terminals L1, L2, and G.

Controller FIC 301 is reverse acting. A decrease in signal from FT 301 will result in an increasing output from the controller. The output of FIC 301 from terminals 15 and 16 of the controller is wired through TS 40, terminals 24 and 25 to cable 30, pair 15. Pair 15 is connected to junction box JB 40, terminals 14 and 15. There it is connected to cable 10 pair 2 to junction box JB 30, terminals 6 and 7. From JB 30 a single pair goes to FY 301A, the I/P. The wire numbers of this pair are connected: FY 301A-1 to the plus terminal and wire FY 301A2 to the negative. FY 301A requires a 20 psi air supply connected to the supply port (S) and the output connects under the diaphragm of the actuator. The control valve FV 301 will fail closed as indicated by the arrow pointing toward the valve body. To show some differences in the details that can be provided on Loop Diagrams, Figure 7-8 is the same loop with some additional, optional items included. Just to be different, wiring is depicted using the alternate ISA symbol of a triple crosshatch rather than the more common dotted line used in Figure 7-7.

The optional items shown on Figure 7-8, Loop Diagram – Electronic Minimum & Optional, include information about the pipe, 1.5" CS Sch 40, and other information:

- Its source is at pump 370 and its termination is in Unit 3.
- There are 20 diameters of straight run piping upstream of the orifice FE 301 and 5 diameters straight run downstream.
- The orifice plate is at elevation 300'0"
- FT 301 is at elevation 290'0"
- There is a three-valve manifold, HV 301, which permits the transmitter to be "zeroed" locally by shutting the upstream valve and opening the

## Figure 7-8: Loop Diagram – Electronic Minimum & Optional

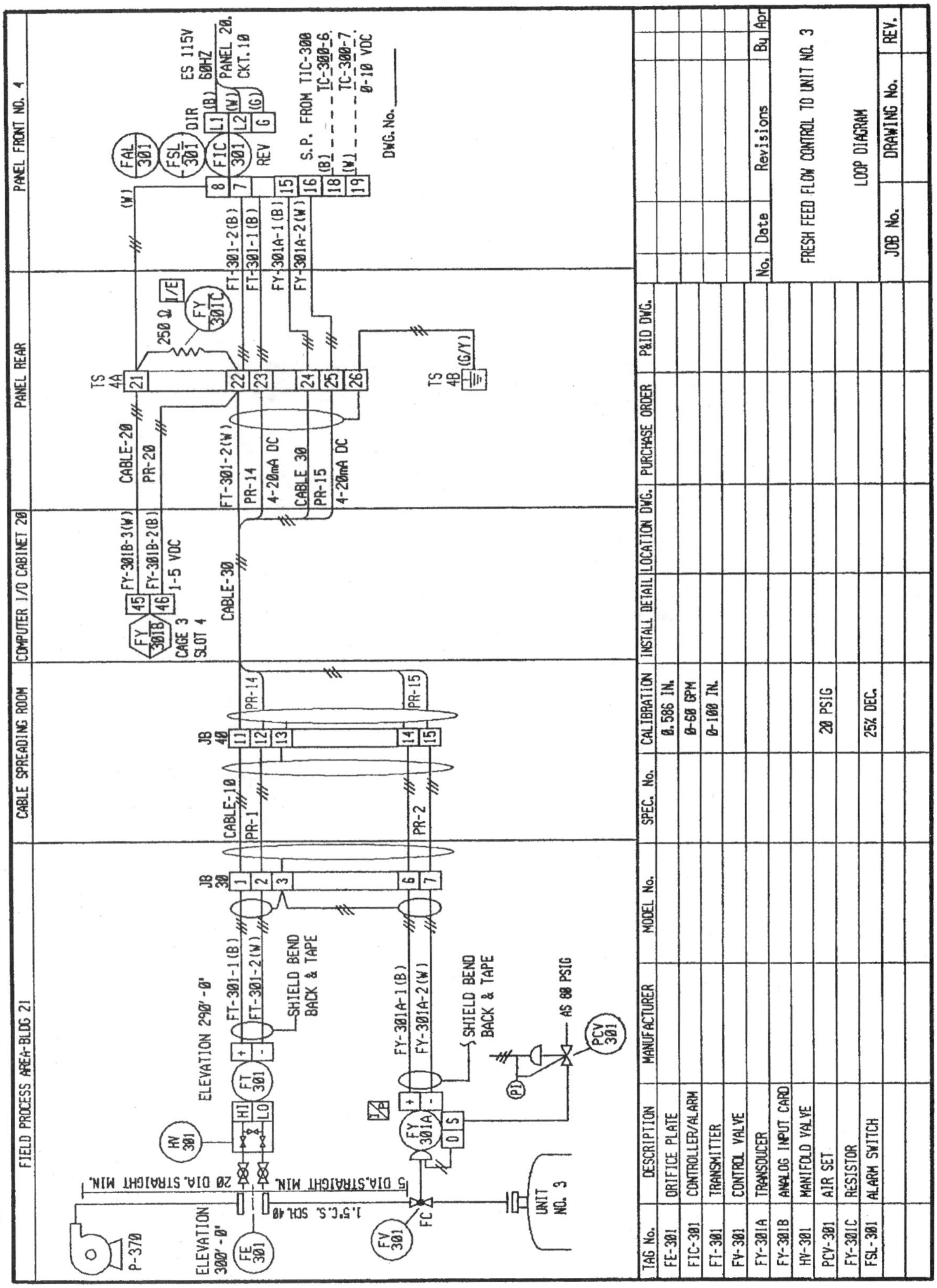


**Loop diagram, electronic control, minimum required items plus optional items.**

*From ISA-5.4*

bypass valve.

- The voltage input of 1-5 volts dc and the location information, Cage 3, Slot 4, is provided for the process control computer.

- Additional tagging has been added for the three valve manifold HV 301, the 250 ohm resistor FY 301C and the pressure regulator for the instrument air supply PCV 301.

- The color of the panel wiring has been specified: "W" for white, "B" for black, and "G/Y" for green with yellow trace.

- Signal levels for the controller input and output are stated as 4-20 mA dc and the computer input 1-5 volts dc. Calling out the signal level may seem a bit obvious or even silly to those of us that have used one analog signal for the past 20 years. Remember though, there are many of our co-workers that are mixing voltage, milliamp and fieldbus signals within their facilities. Calling out the signals may not seem so redundant to those maintenance personnel.

- One large additional feature is the table at the bottom of the Loop Diagram which lists the tag numbers and the following information for all the devices shown on the Loop Diagram.

| Description | Spec number | Location Drawing |
|---|---|---|
| Manufacturer | Calibration | Purchase Order |
| Model Number | Installation Detail | P&ID Drawing |

The above information is a repeat of that shown on other project documents. All this information and more could be included in a project database.

## Loop Diagrams During Design and Construction

Loop Diagrams provide a device for checking the completeness and accuracy of the instrumentation and controls design package. In other words, the process control design is probably complete when all the information needed to complete the Loop Diagram is available. Loop Diagrams are useful during construction, and depending on the other documents prepared for the project, Loop Diagrams may be the primary construction document. Since a proper Loop Diagram is a complete representation of the interconnection of a process control loop, they make a great installation document. Frequently, construction progress is measured by the number of "loops" completed.

One of the last tasks of the instrumentation and controls group prior to start-up is to check out the entire system to verify the devices were installed correctly, and assure the system is ready for operation. These tasks may be accomplished with a loop-by-loop confirmation. These confirmations are called many things: loop checks, process system inspections, construction acceptance checks, or, in our house, "why daddy didn't get home on time". All the necessary information to perform loop checks may be available on other design documents. However, it is much more logical, economical and convenient for the instruments and controls design group to use Loop Diagrams for this task.

The end of construction is usually defined in the construction contract. The term "mechanical completion" may be used. This is a very broad term that needs, but rarely receives, detailed definition. However, a common and perhaps simplistic definition might be: construction is complete, power is turned on, utilities are in operation, but no process fluids have been introduced.

Under a contract with these terms, loop checks might confirm that all the tag number devices are calibrated and interconnected completely and correctly. Loop checks might also confirm the presence and pressure of the instrument air supply, and assure the air is free of dirt and moisture. Electrical power is verified to be at the correct voltage level, properly connected and identified. The owner's operating personnel would probably be involved in operating the utilities and overseeing the power system during loop checks. Process fluids can be simulated, particularly with intelligent or smart field devices.

Other contracts define the end of construction in different terms, and owner personnel might perform the loop checks as part of the pre-startup activity. In any case, for a startup to go smoothly, a complete set of loop checks should be done.

Loop Diagrams are also of value to the operations and maintenance instrumentation and controls group. They are invaluable when troubleshooting a malfunctioning loop, during periodic device recalibrations, preventive maintenance, and during any modification of the instrumentation and controls system.

There is a wide range of opinion on the value of Loop Diagrams. Amazingly, in our experience, there are some owners who believe the cost of developing Loop Diagrams is a needless waste of money. Other owners see the need for Loop Diagrams but believe owner personnel can produce these more effectively, perhaps by the engineers and technicians who will operate and maintain the plant. Still others, and some engineering contractors, use outside help to develop the Loop Diagrams. Some of the major instrumentation and controls suppliers have the capability to develop Loop Diagrams. Some distributed control systems have configuration software that documents the software in a "loop-like" way. There are firms that specialize in developing Loop Diagrams. An increasing number of firms have even made the Loop Diagram and the P&ID the only record drawings for all instrumentation and controls devices – no longer keeping other instrumentation and controls documents in their records.

The use of a common project database rather than the use of other drawings as information sources can simplify the task of developing Loop Diagrams. There are project database management software products available commercially.

Many companies have software developed in-house by dedicated project team members that will produce Loop Diagrams or listings of the same data presented on Loop Diagrams. Some software has the capability of developing and drafting Loop Diagrams directly.

---

**Loop Diagrams – A Look Ahead**

The newer digital automation technologies, such as fieldbus, will call for significant changes in some documents used to define instrumentation and control systems. An article in *Control*, titled *An Engineering and Construction Firm Tackles Foundation Fieldbus*, suggests one possible change is to replace traditional Loop Diagrams with fieldbus segment diagrams.

The article states: "A divide no longer exists between field instrumentation and the control system as it has with DCSs. The control system and the field instrumentation must be engineered together. We call the end result of this combined engineering effort the Plant Automation System (PAS) ....Devices are effectively wired in parallel on a H1 segment, and they share a single port on a process controller's fieldbus communication module. A shielded pair Foundation fieldbus H1 segment communicates at 31.25 Kbps and can address as many as 32 fieldbus devices...Loop diagrams are replaced by fieldbus segment diagrams."[3]

Fieldbus segment diagrams show all of the devices in a single segment and the electrical connection details, including shielding and grounding for all of the devices in the segment. The segment diagrams will serve the same purpose for the PAS as the Loop Diagrams presently do for pneumatic, electronic, and shared display shared control systems.[3]

---

## Summary

In this chapter we have described Loop Diagrams, discussed their use and some of the pros and cons involved. The authors are advocates of developing Loop Diagrams as part of the project deliverables. We believe the project team development of Loop Diagrams forces a last check of the accuracy and completeness of the project design.

If the Loop Diagrams are produced by the instrumentation and controls design group in a timely manner, they are available for loop checks as part of the mechanical completion and for trouble shooting during plant start up. They do constitute a significant expense. To offset this cost, we are convinced that Loop Diagrams for an unknown but significant number of loops will be developed during the life of the plant.

We further believe it is infinitely better to develop all Loop Diagrams in a controlled manner in a comfortable office. Otherwise you might be forced to develop some at 2 a.m. on a rainy night when the plant is shut down because of a troublesome loop, and the plant manager is building up a full head of steam, demanding the plant to restart immediately. Yes, it happens.

---

1. *The Automation, Systems, and Instrumentation Dictionary*, 4th edition (Research Triangle Park, NC: ISA — The Instrumentation, Systems, and Automation Society, 1984), p.299

2. ibid. p.299

3. Andrew Houghton and David Hyde, *An Engineering and Construction Firm Tackles Foundation Fieldbus*, *Control*, (Itasca, IL: Putman Media Co., February 2001)

# Installation Details and Location Plans

## Installation Details

Installation Details define the requirements to correctly install an instrumentation and control component – an instrument or control valve. These details establish requirements for supporting an instrument and mechanically connecting it with the process. Unlike the schematic representation of the Loop Diagram, Installation Details refers to the actual physical requirements of an installation; it may detail the very nuts and bolts needed.

Installation Details show where the connection will be made on the piping, tank, vessel, chest or duct. The relative position of the instrument to the process connection and to the floor is provided. When necessary, Installation Details are also used to show how the components are insulated, winterized, or otherwise isolated from high or low temperatures.

There is no ISA standard that defines requirements for Installation Details, nor can a reference to Installation Details be found in the ISA dictionary. However, that does not mean there is a lack of information on installing I&C devices. Libraries of Installation Details are established and maintained by plant owners, maintenance technicians, instrument manufacturers, engineering contractors, instrument installation contractors and even by individuals. Of course, the libraries all differ in detail and nomenclature. Furthermore, most of the owners of Installation Details are convinced their set will give the best results when used. While there is no standard definition for Installation Details, many follow these general guidelines:

- Installation Details will be used in the field, so an 8 ½" by 11" drawing is the optimal size. It is easy to carry and to reference during installation. Some Installation Details are designed in 11" x 17" format but are reduced to the smaller size for handling.

## Bills of Material

- For control system documentation, bills of material, or material lists, are found only on Installation Details. Material lists, or bills of material, are tables with item numbers that correspond to identification on the body of the drawing. The tables also include quantity counts, a description of the item, and often a manufacturer and a part number. The information shown is often sufficient to purchase the components. For Installation Details, tagged devices do not appear on the bills of material. Exact counts for tubing lengths may not be provided since the required length will vary. "A/R" for "as required" may be in the quantity column. Some Installation Details will call out a conservative but "standard" length for tubing to support the development of material take offs. Usually material lists are included in the scope of the Installation Detail. The bills of material may appear on the drawing itself. Alternatively, an 8 ½" by 11" drawing may be used for the Installation Detail and a separate 8 ½" by 11" text sheet is attached for the material list. This approach was particularly common prior to computer-aided design and drafting (CADD). The material list was typed on a separate sheet and attached to the hand drawn Installation Detail.

- The Installation Detail depicts only the installation of a single device – unlike the Loop Drawing, which shows all the devices associated with the loop number. Thus, your facility may use separate Installation Details for an I/P converter and for a control valve. Therefore, there could be three details: one for a separately mounted I/P, one for a control valve without an I/P, and a third for an I/P mounted on a control valve. Figure 8-2 shows a very common Installation Detail depicting installation requirements for a remote mounted differential pressure transmitter in gas service, drawing ID-101.

- There should be some link or cross reference in your documents between a device tag number and the associated Installation Detail. Two common approaches are to call out the Installation Detail on the Instrument List; the other is to issue a unique Installation Detail for each device tag number. In our example with the "remote mounted differential pressure transmitter in gas service" instrument, drawing ID-101, the first method would yield a reference to "ID-101" in the Instrument List for every dp cell of that type. The second method of using a unique drawing for each tag number would yield many individual copies of the same drawing ID-101, with each copy differing only by the referenced instrument tag number. There are advantages to either approach. The former calls for fewer drawings and might make drawing revisions a bit simpler to issue. It does, however, assume the concerned parties are comfortable with using the Instrument List to direct work performed by the installation contractor. The latter method is less prone to misunderstandings, but there must be a system, budget and willingness to handle the increased number of drawings.

- Usually separate Installation Details are used to show how the device is mounted, how it is connected to the process and how the electrical connection is made. Therefore, a single tag marked instrument will require more than one detail to show the installation.

- Project managers or facility owners occasionally request that all Installation Details for a single tag-marked device be included in one drawing. This unwieldy requirement may lessen the standardization advantage to using Installation Details, since many more individual drawings will have to be prepared. Also, and possibly more importantly, it is difficult to fit all the mounting, installation and wiring connection information on a single sheet. One solution is to reference mounting and wiring details on the primary Installation Detail so only the primary detail needs to be issued or referenced to the device tag number. Nonetheless, complex or special instrument installations may require preparing a few dedicated and unique details, but this should not occur frequently.

- It is very common for facilities, projects, construction contracts, or even tradition to call for inclusion of a material list on the Installation Detail. The material list, also known as bills of material, spells out the manufacturer, model number, material and quantity of the components needed for a single installation. Facilities with different process driven material requirements might use a written instrument piping material specification to support the Installation Details rather than attempting to cover all the material combinations on the Installation Details. The different processes would be cross-referenced to the appropriate materials listed in the specification. This is similar to the system used for piping specifications. On less complex projects, the instrument piping specification may simplify the material callouts by standardizing on the "worst case" material specification, probably 316L low carbon stainless steel tubing and stainless steel fittings.

- The use of a material list allows the installer to make easy and accurate material take-offs used for preparing material requisitions and purchase. The material list also enables warehouse personnel to pull and bag material necessary for an instrument installation. The instrument installer can then pick up a bag containing all the material necessary to install a particular device. And you thought one-stop shopping only applied to the supermarket!

- Many plants and construction contractors have work rules assigning a responsible craft for specific work tasks, notably the mechanical installation, the pipe fitters, and the electrical installation, the electricians. To prevent conflicts, the industry generally tends to avoid showing both electrical and mechanical installation details on the same drawing. Furthermore, having a third separate detail for mounting the device allows assigning that work to either contractor, depending on local work agreements.

## Types of Installation Details

There are three general types of Installation Details:

*(Note: the Type 1, 2, and 3 classification used below is merely for clarification. It is not a term used elsewhere.)*

**Type 1, Inter-discipline design:** Developed to transfer information between the design teams, notably between the I&C designer and the piping designer. These details can be informal sketches or formal drawings. These design coordination documents will show installation requirements for components installed in piping. Layout details are not useful to provide to the installation contractor, since by the time the installation contractor is working, the piping design is complete, the spool drawings issued, and the piping fabricated. Once developed, these details should become part of the company or plant design standard for use on all subsequent projects. Some typical items covered might include:

- Upstream and downstream straight run pipe length rules for flow elements – X diameters upstream and Y diameters down, depending on the piping configuration.

- Magnetic flow meter is always flooded, or the flow is vertically upward.

- Piping connections for thermowells; small pipe sizes will swage to 3" or 4", or make use of elbows and tees.

- "Root valve" requirements for pressure connections, perhaps a 6" pipe nipple ending with ½" gate valve.

- Allowable location of taps in piping; top of pipe, side, bottom, anywhere on an arc from 10 degrees to 350 degrees, etc.

- Flange mounted level transmitter connection – size, flange rating, who furnishes isolation valve, etc.

**Type 2 – Mechanical Installation**: Shows how the I&C device is to be mounted and connected to the process. These Installation Details are usually the responsibility of the I&C design team, and they cover the installation from the first isolation valve at the process, called the root valve, to the instrument itself. It is on these details that the tubing requirements are established – including the slope, tubing size, and relative position of the transmitter – to the process connection, use of flexible hose for air and signal connections, to a control valve, etc.

**Type 3 – Electrical Installation**: Shows how the electrical connections are made. Either the electrical installation details are the responsibility of the electrical design team or the electrical team is integrally involved in their development. The electrical group will ensure the instrument installation is in concert with the balance of the electrical work, and also ensure the installations meet local and national electrical codes. Information included in these details include the use of flex conduit, cable grips with open cable drops to instruments, handling pig tail wiring for instruments, junction box mounting, and, most importantly, conduit and wire tagging requirements.

### Kept in Library

Installation Details often reside in a library of standard details for your facility. The library will build over time as new devices are specified that offer new installation requirements. The details are the responsibility of the I&C design team or maintenance group. In a perfect world, your facility should have Installation Details covering every type of instrument or control device in your facility. That being said, sometimes the manufacturer's Installation Details will be used instead of a project-generated detail. However, a bit of prudence must be used when using a manufacturer's details by themselves. The manufacturer's detail may not cover the complete installation or, more frequently, the manufacturer's detail will sometimes confuse an installer by providing generic information or information on options not used for your installation.

A complete set of Installation Details is generally developed as part of a project design by the I&C and electrical design teams. Of course, there are exceptions to this approach.

On occasion, an operating facility will subcontract the instrument and controls installation to an instrument installation contractor. The project team should ensure the contractor's details will meet the site standards. This verification should, of course, take place prior to signing the installation contract, to prevent embarrassing and expensive change orders. Some instrument manufacturers also have the capability to undertake some or all of an installation contract. In both of these instances, developing or updating the site Installation Details might be included as part of the installation contract.

There are certainly more variations on Installation Details than have been covered here. We have seen increasing use of digital photos to present preferred installations, complete with overwritten notes and leader arrows to significant points. At least one oil industry plant does without any Installation Details for all in-house projects. They add some installation information to the Loop Diagram and use this for installation directions. Their installers have the skill necessary for a satisfactory installation.

## Needed for Bidding

Installation Details are an important bid item when soliciting construction quotes. Some projects have dealt with contract change orders because, during the bid process, an owner provided out-of-date or non-representative Installation Details. The construction team later discovered the actual installation requirements were much different from the basis for their quote.

Figure 8-1 is an example of a Type 1 Installation Detail developed to transfer information from one discipline to another during the project design phase. These details may be developed as company or project standards, perhaps long before the present project started.

Figure 8-1 shows how thermowells are to be installed in piping. Thermowells installed in this manner will permit accurate fluid process temperature readings. These installations provide a self-flushing thermowell installation to mini-

**Figure 8-1: Installation Detail, Type 1 – Thermowell**

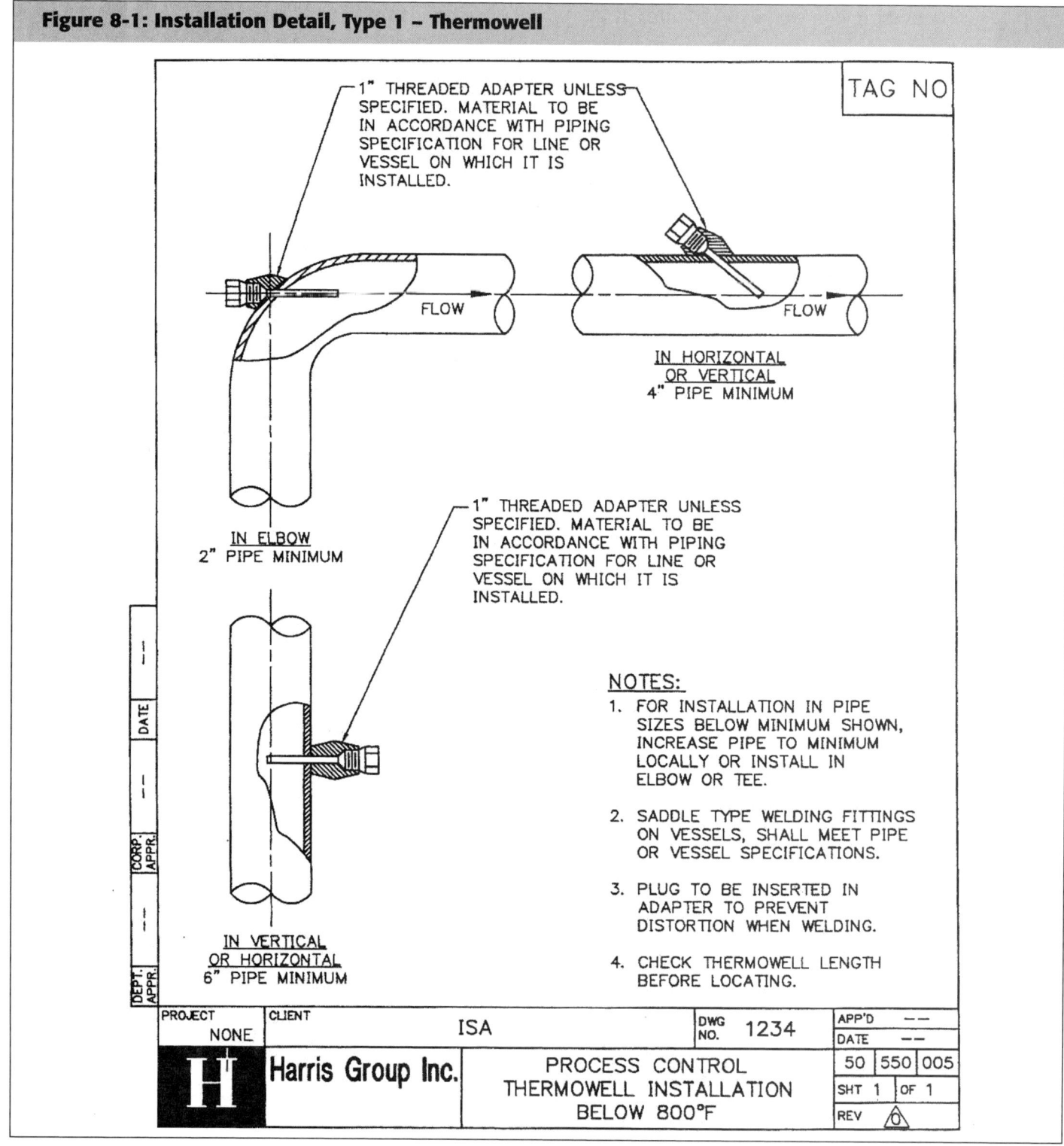

1" THREADED ADAPTER UNLESS SPECIFIED. MATERIAL TO BE IN ACCORDANCE WITH PIPING SPECIFICATION FOR LINE OR VESSEL ON WHICH IT IS INSTALLED.

TAG NO

FLOW

FLOW

IN HORIZONTAL OR VERTICAL
4" PIPE MINIMUM

IN ELBOW
2" PIPE MINIMUM

1" THREADED ADAPTER UNLESS SPECIFIED. MATERIAL TO BE IN ACCORDANCE WITH PIPING SPECIFICATION FOR LINE OR VESSEL ON WHICH IT IS INSTALLED.

NOTES:

1. FOR INSTALLATION IN PIPE SIZES BELOW MINIMUM SHOWN, INCREASE PIPE TO MINIMUM LOCALLY OR INSTALL IN ELBOW OR TEE.

2. SADDLE TYPE WELDING FITTINGS ON VESSELS, SHALL MEET PIPE OR VESSEL SPECIFICATIONS.

3. PLUG TO BE INSERTED IN ADAPTER TO PREVENT DISTORTION WHEN WELDING.

4. CHECK THERMOWELL LENGTH BEFORE LOCATING.

IN VERTICAL OR HORIZONTAL
6" PIPE MINIMUM

DATE / CORP. APPR. / DEPT. APPR.

| PROJECT | CLIENT | | DWG NO. | APP'D | — — |
|---------|--------|--|---------|-------|-----|
| NONE | ISA | | 1234 | DATE | — — |

Harris Group Inc. | PROCESS CONTROL THERMOWELL INSTALLATION BELOW 800°F | 50 550 005 / SHT 1 OF 1 / REV △

mize buildup on the thermowell protrusion in the process flow; that is, it keeps junk and trash off the well. This buildup will, of course, affect the accuracy of the temperature reading.

Figure 8-2 is an example of a Type 2 Installation Detail. Type 2 details are usually defined, developed or selected from an existing library of standard details. They are the responsibility of the I&C design team and are generated specifi-

**Figure 8-2: Installation Detail, Type 2 – Flow Transmitter**

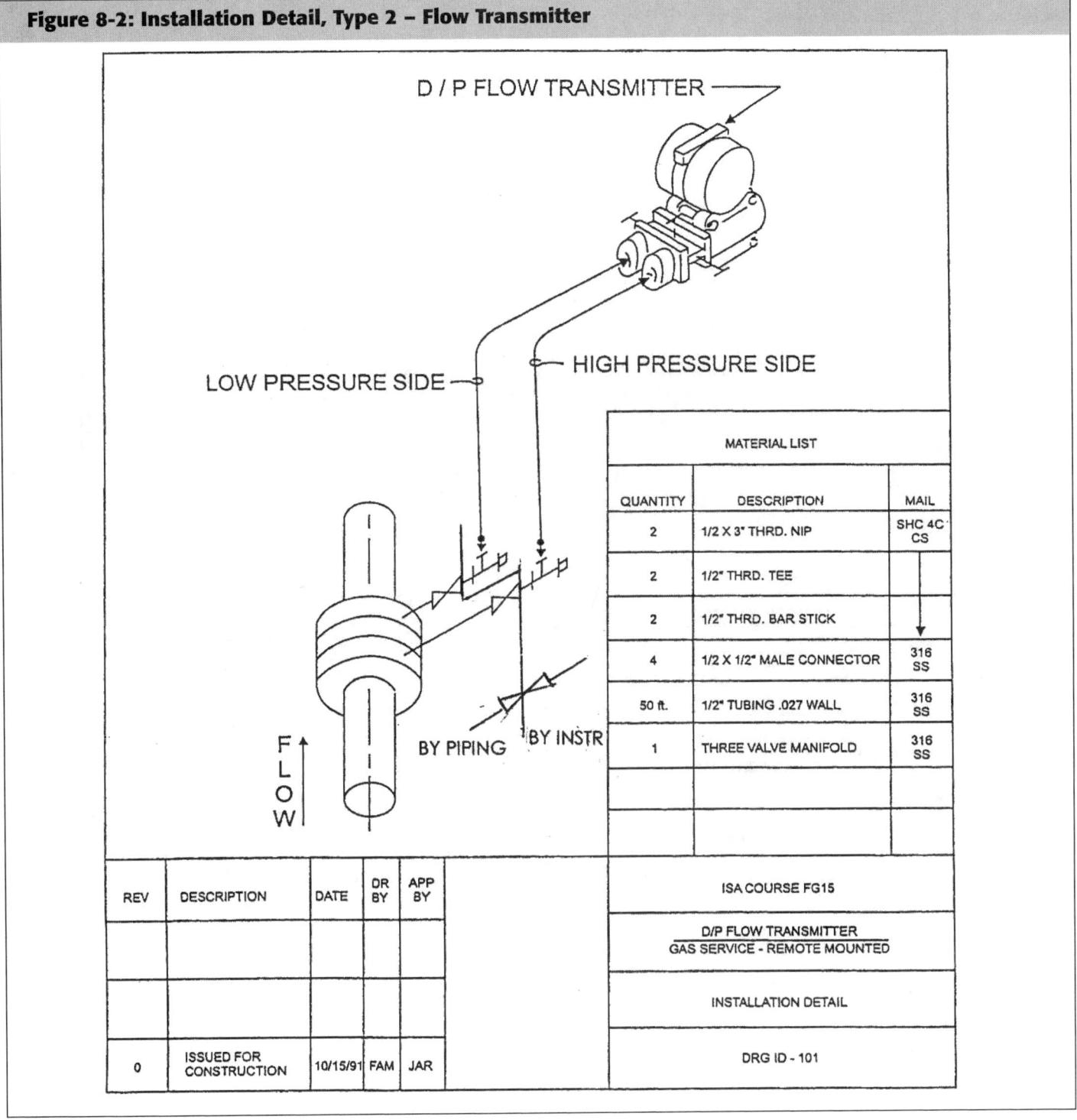

D / P FLOW TRANSMITTER

LOW PRESSURE SIDE — HIGH PRESSURE SIDE

| | MATERIAL LIST | |
|---|---|---|
| QUANTITY | DESCRIPTION | MAIL |
| 2 | 1/2 X 3" THRD. NIP | SHC 4C CS |
| 2 | 1/2" THRD. TEE | |
| 2 | 1/2" THRD. BAR STICK | |
| 4 | 1/2 X 1/2" MALE CONNECTOR | 316 SS |
| 50 ft. | 1/2" TUBING .027 WALL | 316 SS |
| 1 | THREE VALVE MANIFOLD | 316 SS |

BY PIPING   BY INSTR   FLOW

| REV | DESCRIPTION | DATE | DR BY | APP BY | | ISA COURSE FG15 |
|---|---|---|---|---|---|---|
| | | | | | | D/P FLOW TRANSMITTER<br>GAS SERVICE - REMOTE MOUNTED |
| | | | | | | INSTALLATION DETAIL |
| 0 | ISSUED FOR CONSTRUCTION | 10/15/91 | FAM | JAR | | DRG ID - 101 |

cally for all devices in the project. They are issued to installers for use during construction.

Figure 8-3 is an example of a Type 3 Installation Detail. It shows the general-purpose conduit installation for an instrument or control component. In this instance, the detail was developed and is maintained by the electrical group. Note the drawing number ELEC-001.

**Figure 8-3: Installation Detail, Type 3 – Conduit**

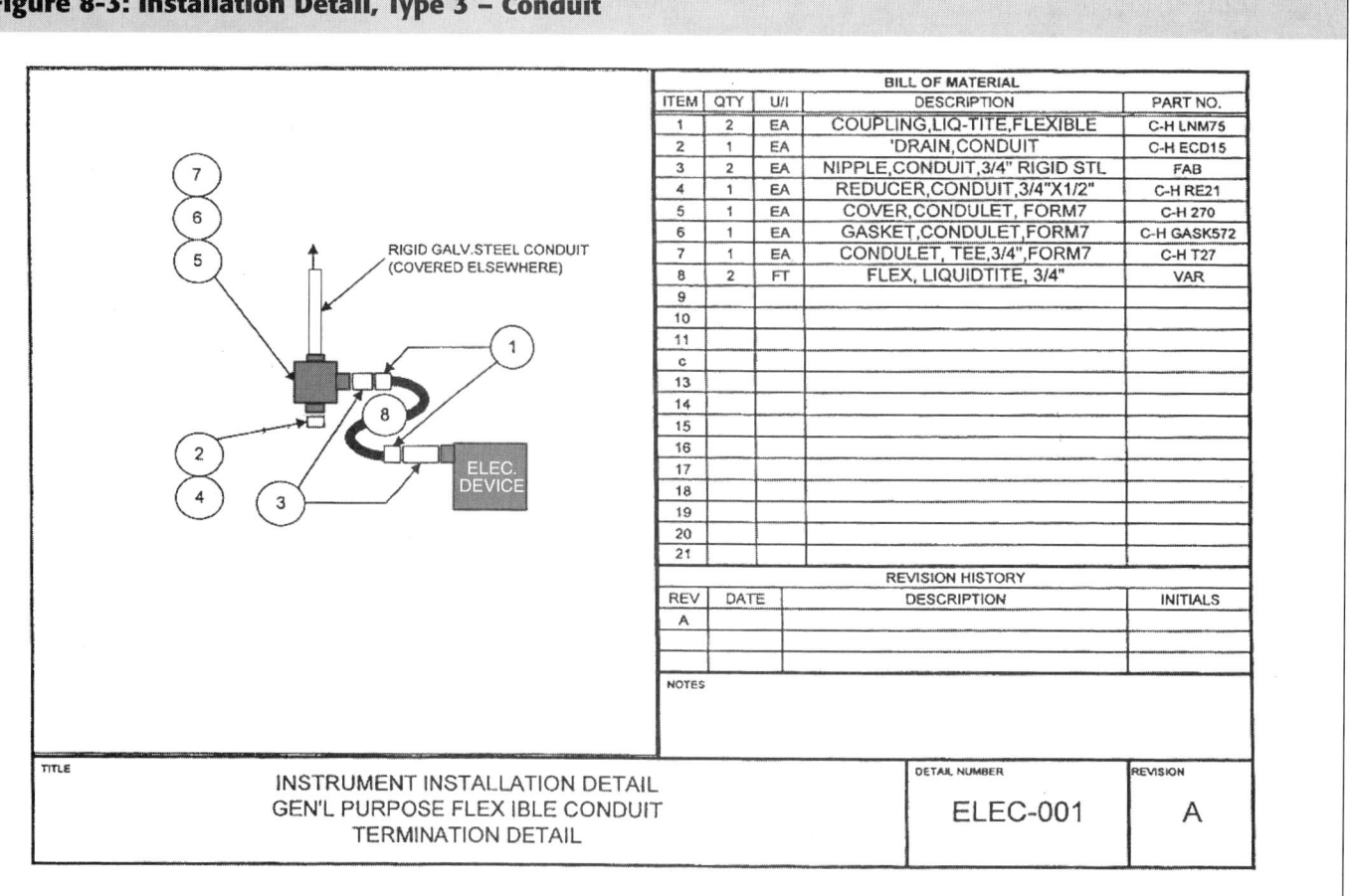

| BILL OF MATERIAL | | | | |
|---|---|---|---|---|
| ITEM | QTY | U/I | DESCRIPTION | PART NO. |
| 1 | 2 | EA | COUPLING,LIQ-TITE,FLEXIBLE | C-H LNM75 |
| 2 | 1 | EA | 'DRAIN,CONDUIT | C-H ECD15 |
| 3 | 2 | EA | NIPPLE,CONDUIT,3/4" RIGID STL | FAB |
| 4 | 1 | EA | REDUCER,CONDUIT,3/4"X1/2" | C-H RE21 |
| 5 | 1 | EA | COVER,CONDULET, FORM7 | C-H 270 |
| 6 | 1 | EA | GASKET,CONDULET,FORM7 | C-H GASK572 |
| 7 | 1 | EA | CONDULET, TEE,3/4",FORM7 | C-H T27 |
| 8 | 2 | FT | FLEX, LIQUIDTITE, 3/4" | VAR |
| 9 | | | | |
| 10 | | | | |
| 11 | | | | |
| c | | | | |
| 13 | | | | |
| 14 | | | | |
| 15 | | | | |
| 16 | | | | |
| 17 | | | | |
| 18 | | | | |
| 19 | | | | |
| 20 | | | | |
| 21 | | | | |

| REVISION HISTORY | | | |
|---|---|---|---|
| REV | DATE | DESCRIPTION | INITIALS |
| A | | | |
| | | | |
| | | | |

NOTES

| TITLE | DETAIL NUMBER | REVISION |
|---|---|---|
| INSTRUMENT INSTALLATION DETAIL GEN'L PURPOSE FLEX IBLE CONDUIT TERMINATION DETAIL | ELEC-001 | A |

## Location Plans

There is no ISA standard that defines a Location Plan. The ISA dictionary has no reference to a Location Plan. Our definition, for the purpose of this book is: Location Plans show the field instruments and control devices – the transmitters, control valves, transducers, local panels, junction boxes, termination points for field I/O, etc. A general rule would be to show anything the I&C group "owns", meaning devices the I&C group installs or interconnects with wiring or tubing.

All field instruments are shown using simple notation, perhaps only by the tag number and, if your standards call for it, the elevation of the device. The ubiquitous ISA instrument bubble is frequently used. The bubble is less effective for densely packed drawings; it uses up a lot of space. When the Location Plans have to show many devices in small space, one effective solution is to use a simple text string for the tag. In this case, the I&C device is located with a solid symbol like a small square. A leader line connects the solid symbol to the text string.

Location Plans use, as their base, orthographic scale drawings of the facility. Normally, these scale drawing backgrounds are developed by other design groups, typically by the architectural, mechanical, or piping groups. This works well from a scheduling standpoint as the other groups will be, or should be, essentially complete with their design efforts by the time the instruments are located. Instrument locations are sometimes added to the electrical plans, and sometimes the I&C group will prepare its own. The quantity of instruments and the density of the electrical plans will dictate which system is used. There must be close coordination between the two groups to make efficient use of the electrical "backbone" – the cable tray and conduit runs. Adding a 24v dc signal tray next to a 120v ac low voltage power tray is easier than finding room for, and supporting, the two trays independently. All parties should agree to the content and scope of the Location Plans before design commences, since there are many possible variations.

Location Plans work well using a fadeout piping drawing or, better yet, fadeout equipment layout drawing. In a fadeout drawing, details that are not necessary for the Location Plan are shown using very light lines (faded). This is quite easy to accomplish using CADD or, with a clever reproduction facility, printing onto Mylar drawing stock. The information added for instrument device locations is drawn in normal line weight so it will show clearly when contrasted to the fadeout background.

Figure 8-4 Location Plan, Approach A, is typical of one type of Location Plan. This approach locates devices orthographically and, sometimes, the elevations

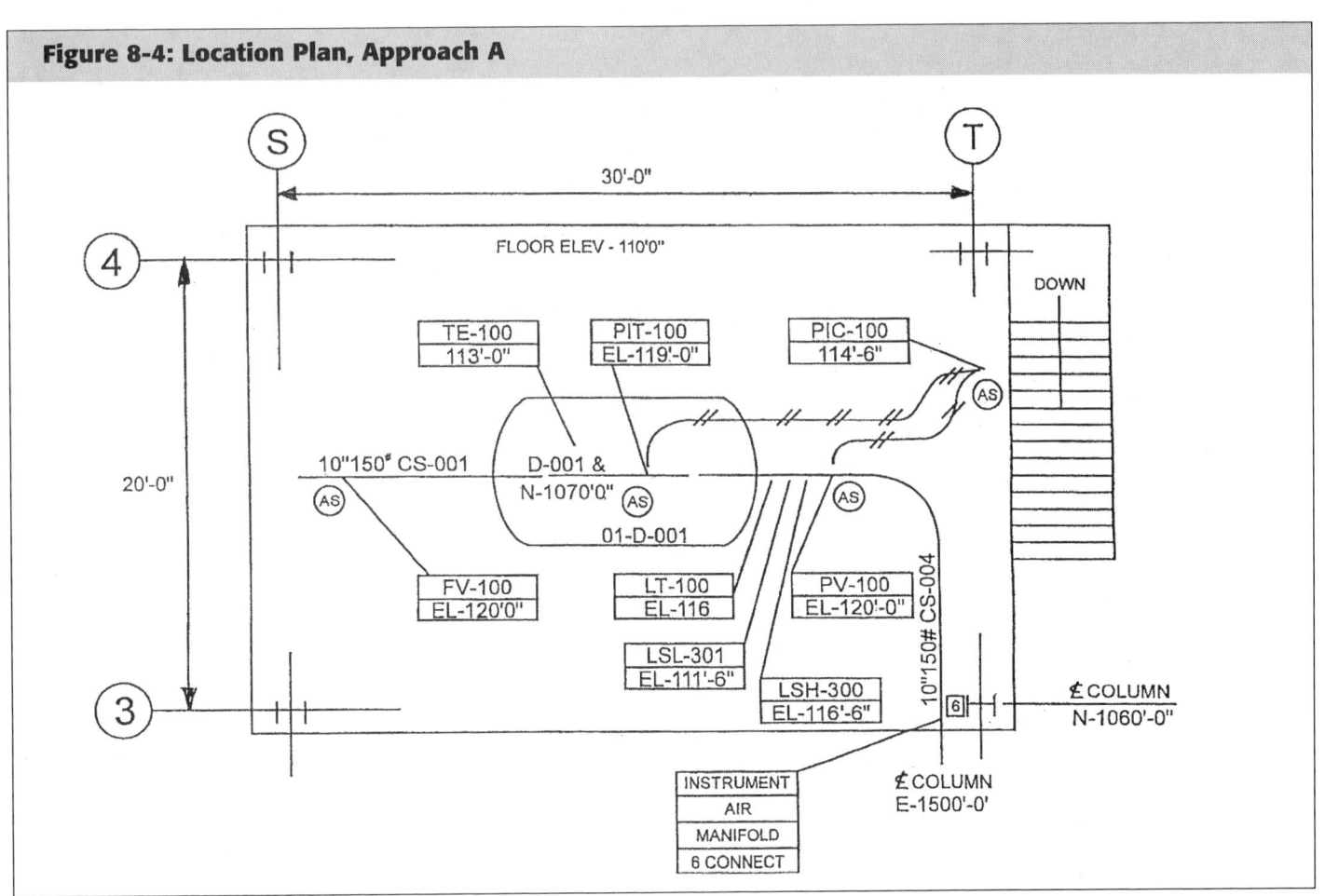

**Figure 8-4: Location Plan, Approach A**

are indicated. This example uses some of the devices as shown on the example P&ID from Chapter 2, Figure 2-21 (page 45). The plan drawing shows the location of the instrument air manifold and defines the transition point between the instrument air distribution header and an instrument air branch header. The transition is important since it will be the boundary between work by the piping group, the instrument air header, and the instrument mechanical contractor, the instrument air branch piping.

The plan also indicates which devices require connecting to an instrument air supply and shows in sketch format the necessary pneumatic transmission lines for loop PIC-100. The Location Plan uses the plant identification grid, and rows and columns, including the dimension callouts. Therefore, grid locations for all the devices may be developed from this Location Plan. Location Plans will be useful when assigning individual instruments to I/O panels; the designers can assign devices to the nearest panel.

## Grid References

Having accurate instrument locations can be very helpful to maintenance personnel, to the installation contractor and to the design staff. In fact, having locations shown in the Instrument List is a requirement at some facilities. This location information can be time consuming to acquire, so it can be expensive information. Making matters worse, for those who are working on a construction project, defining locations normally occurs relatively late in the project, when budgets are tight.

The question always arises: do you call out the nearest column intersection when defining a device location? Define "*nearest*", because sometimes it isn't obvious which column intersection is closest. Using this method, the device could be in any direction relative to the intersection, but within half the distance to the next column. Alternatively, do you take the time to estimate the distance to the two nearest columns for a more exact map? In our opinion, an exact location in the list is rarely important. It is sufficient to simply identify the location within 20 feet or so; therefore, the column intersection is usually sufficient. There is little need to subdivide the column callouts. Location at column DD-21 is sufficient, rather than fine tuning the location to DD.2-21.4.

Next, you need to decide which intersection to use. We have had the most success with establishing early in the project that the "upper left" intersection, or more technically, using the northwest intersection as the callout. This way, the information is easy to figure out, easy to maintain and everyone knows how to use the information, since the "northwest" intersection rule is easy to remember.

The Location Plan for this project does not show routing of the air supply tubing. The tubing for this project will be field-run and supported by instrument installation personnel. Many Location Plans do not show conduit runs. Many industrial facilities establish the conduit routing in the field so there is little value in adding it to the drawing. It may be useful to show major electrical raceway like multiple conduit runs and cable trays above a certain size, perhaps greater than 6". Then the locations can be coordinated with other disciplines. It is arguably better to have one person or team run all the raceway design, so it is normally more efficient to include the instrumentation tray and main conduit runs in the electrical design package.

The example Location Plan shows the location of TE-100, a thermocouple connected to a control board mounted indicator, TI-100. On this project, the Location Plan does not show the location for PI-1, since its location will be shown on piping drawings. No location is provided on the Location Plan for TI-1 since it will be shown on the vessel drawing of 01-D-001, the KO drum. The rules for locating devices must, as you can see, be worked out between the design disciplines and the owner before the design commences. Actually, it should be decided before the various design budgets are established.

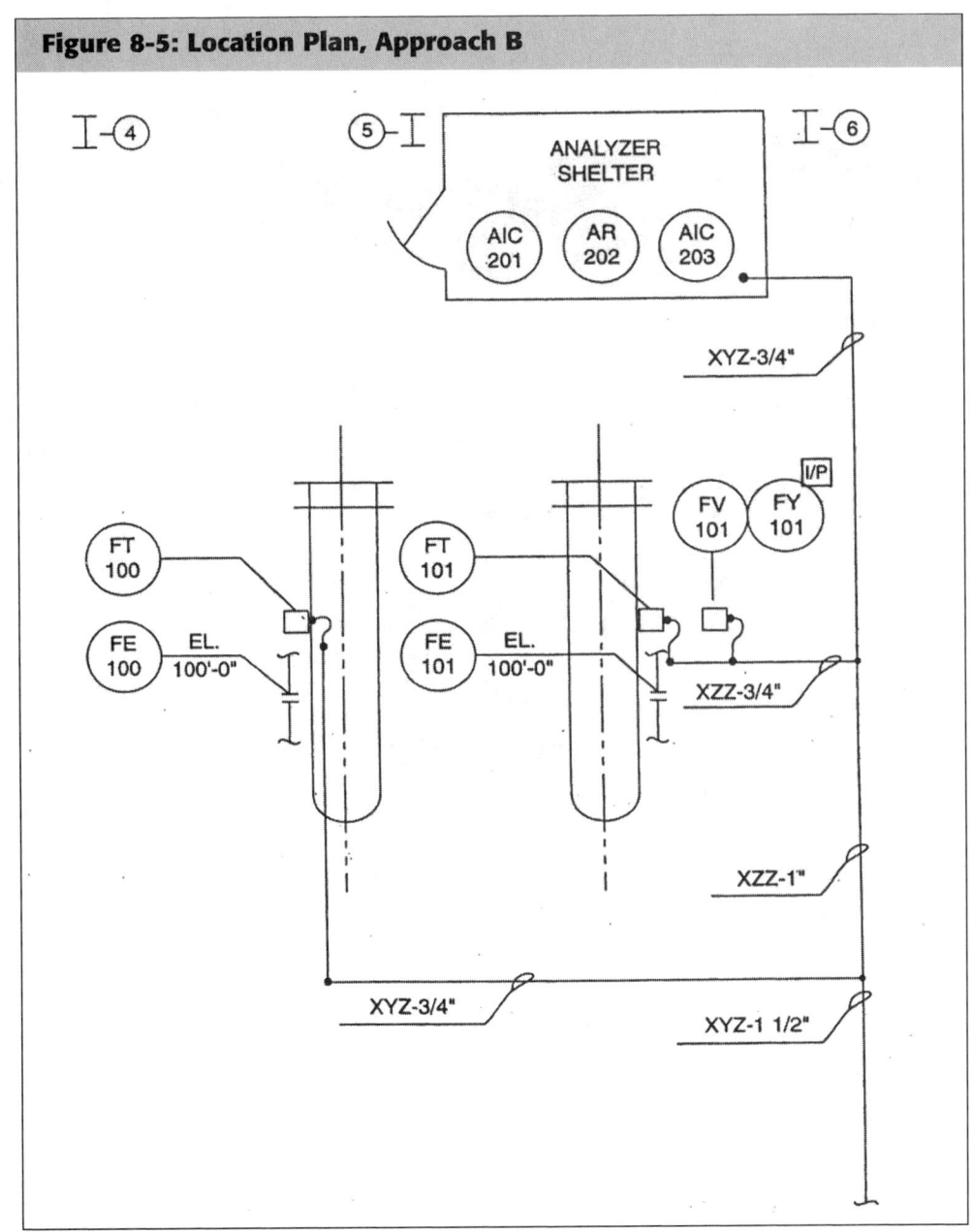

**Figure 8-5: Location Plan, Approach B**

*From Control System Documentation, by Raymond Mulley, ISA 1994*

Figure 8-5, Location Plan Approach B, is drawn to a different set of project criteria. The P&ID on which it is based is not included in this book. This Location Plan shows the relevant column line, but not enough information is included to develop grid locations. The Location Plan uses standard "bubbles" or "balloons" and tag marks to identify the various devices. It shows the conduit size and layout for the electronic transmission system. Instrument and valve elevations are shown only when they differ from a project standard of 4' 6" above the platform height. No information on instrument air supply or distribution is included.

**Figure 8-6: Location Plan, Approach C**

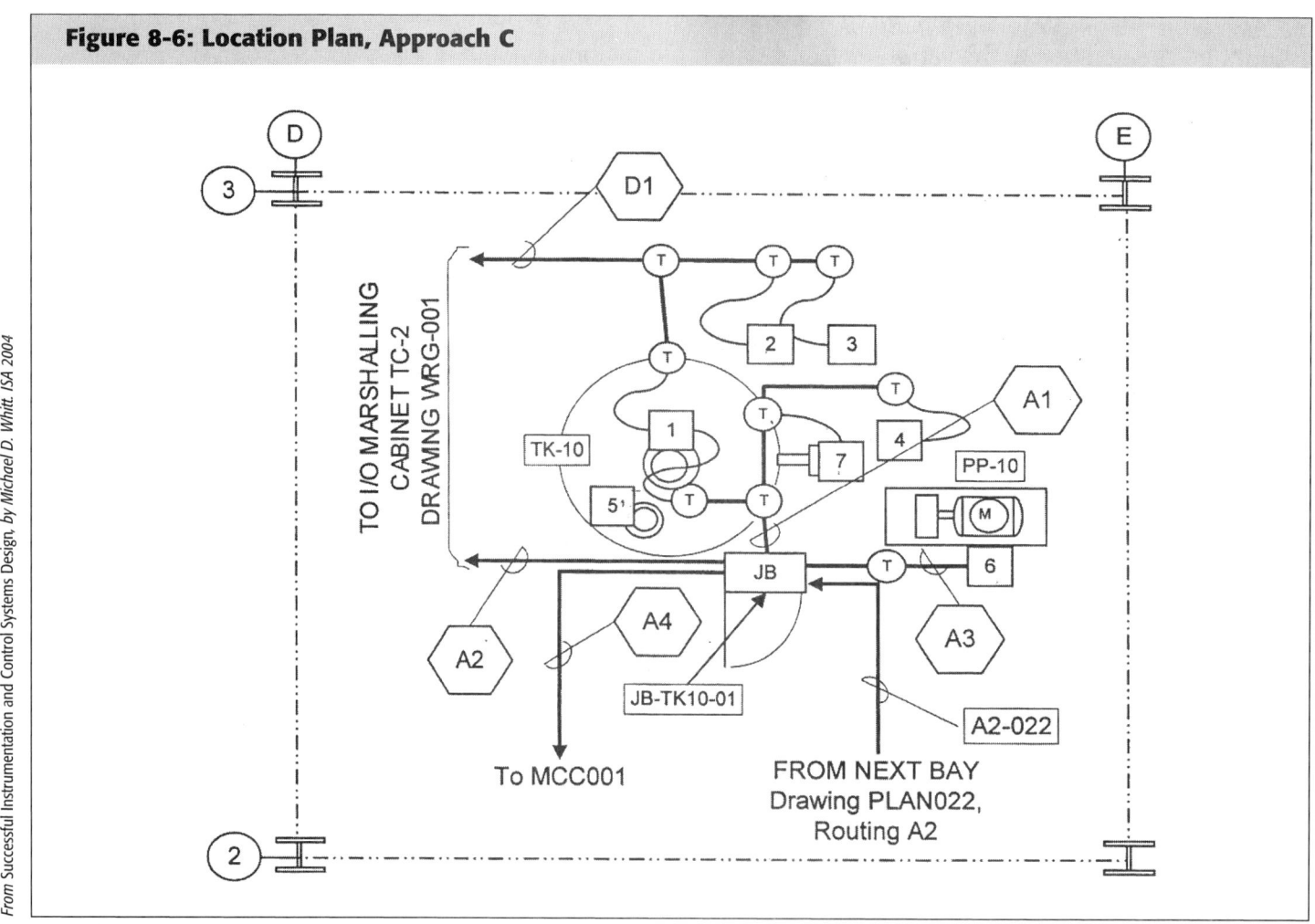

Figure 8-6, Location Plan, Approach C, shows a Location Plan in development. All details are not included. The Location Plan differs substantially from Location Plan A & B. This Location Plan and the following description are taken from *Successful Instrumentation Control System Design*, by Michael D. Whitt.[1]

"Instrument location plan drawings depict the location and identity of instrumentation and control equipment and provide conduit routing and cabling information. The location plan is typically spawned from an equipment arrangement or rendered from a CADD three-dimensional piping arrangement. Once the background is completed showing the equipment arranged on the floor and enough steel detailed to indicate the location within the plant, the I&C designer layers on the instruments and shows conduit routings and contents. Junction boxes are depicted and cable routing schedules are produced. The graphics presentation can differ from one customer to the next depending on standards.... This drawing...shows our TK-10 product tank, a typical vessel with a level control system, fixed-speed discharge pump PP-10, and a recirculation system. The pump has a HOA switch and a contact closure to provide pump status to the computer. Instrument station boxes represent the instruments. The instrument station concepts lets one box represent several instruments providing the instruments are physically grouped. In our simplified example, one box represents one device. However, that device may generate more than one signal.

"Conduit bodies are depicted with a circled letter. The letter indicates the general type of conduit fitting as denoting by the Crouse-Hinds nomenclature. The letter T indicates a tee-type fitting that could be the Crouse-Hinds T- style body or the explosion-proof GUAT-type fitting depending on the area classification....

"Rigid conduit then carries the cabling to its destination. A component schedule describes the instruments represented by the station boxes and details the cabling that feeds them. Notice that items 4 and 5 each have two conduit connections, one for AC power and one for DC signal. Ultimately the conduit size is determined and layered onto the drawing or listed in a cable schedule."

## Detail Minimized

The basic Location Plan shows a minimum amount of detail. Tables are developed and added to the Location Plan as additional information is needed.

The I&C devices are shown as numbered squares, 1 through 7. On a separate component schedule (not shown) item 1 is identified as a level transmitter, LT-TK10-10, and a limit switch, LSH-TK10-10.

Conduits are shown schematically and identified by lettered and numbered hexagons, D1 for dc circuits, and A1, A2, A3, and A4 for ac circuits. The cable origins and destinations are indicated.

A junction box is shown as a lettered (JB) rectangle with further identification shown with an adjacent rectangle (JB-TK10-01). In a recent discussion, I asked a manager from an I&C design firm how his company handles Installation Details and Location Plans. He stated that they rarely, or never, provide Installation Details as part of their deliverables because their clients do not need or want them. His firm does produce Location Plans that differ from approach A, B, or C. Their Location Plans show the location of all tag-marked devices and their elevation on a background outline of major equipment with no interconnection information. Their clients use installation contractors or in-house installers who develop sufficient installation experience, coupled with information from other project documents, to install and connect the devices.

Location Plans are not always included in design deliverables. If a detailed model is built or developed by graphical means for all or part of a plant, the Location Plan may be unnecessary. Other design documents and even the layout of the facility may preclude the need to define device locations and interconnection information. All this being said, it is normally very helpful to provide some form of an overall view of the device layout during construction. It assists field construction supervisors with setting crew size and work schedules, and confirms the availability of installation material.

At least one oil industry facility no longer includes Location Plans and Installation Details in their product design. Their design group augments the Loop Diagrams with additional connection information and uses knowledgeable installers – either in-house technicians or contractors – to install the devices.

## Summary

To summarize, some I&C design packages include Installation Details and Location Plans that locate instruments and show instrument air details but do not show any conduit layouts. Some design packages include the location and elevation of pressure gauges and thermometers and others do not. Some design packages include a Location Plan that shows the locations and elevations of only the devices that require an electrical/electronic connection. Still other design packages include a Location Plan that shows location and elevation of all tag-marked devices, but do not show any instrument air detail or conduit information.

The content of your drawing set must meet your local requirements. There is no universally correct answer, since each facility operates under diverse physical layout, operating, maintenance, design, construction and even regulatory conditions.

Recently I expressed the belief that the Location Plans slowly fall into inaccuracy or even disappear once construction was complete and the plant was operating. Operating personnel, on hearing this statement, corrected me. To them the Location Plan was one of the most valuable I&C documents they had. Since the alternative would be to find a faulty field device by searching through many, many drawings, they valued the Location Plan as a time and money saving document.

---

1.  Whitt, Michael D. *Successful Instrumentation Control System Design* (Research Triangle Park, NC: ISA – The Instrumentation, Systems, and Automation Society, 2004) page 258.

## CHAPTER NINE

# Drawings, Title Blocks & Revisions

The medium upon which we draw has changed much during the past 30 years. What was once ink on linen is now bytes on magnetic media. Drafting boards are now desktop computers. Despite these changes, some features remain – for example, the template of a drawing appears pretty much the same as it did in the past. Several features characterize a template:

- Drawing Size
- Border
- Title Block
- Revision History
- References and Notes

A template is what the drawing looks like before you write on it – for instance, when you pulled it out of the drawer in the drafting room. Today, your company's computer-aided drafting (CAD) program may generate a template automatically.

## Drawing Size

Drawing sizes in the United States are defined in ANSI/ASME Y14.1–1995 (R2002), Decimal Inch Drawing Sheet Size and Format. In it you will find the dimensions and features that comprise an effective drawing. The market also establishes sheet size itself; only certain sizes are readily available. The standard ANSI drawing sizes are identified with a single letter. A summary of the available sizes used in engineering is given in Figure 9-1.

### Figure 9-1: ANSI Document Sizes for Engineering Drawings

|  | ANSI Letter | Size | Margin | Typical Uses |
|---|---|---|---|---|
| **Notebook Size** | A | 8-1/2" x 11" | H-0.38"<br>V-0.25" | Specification Forms<br>Installation Details |
|  | B | 11" x 17" | H-0.38"<br>V-0.62" | Loop Diagrams<br>Installation Details |
| **Desktop Size** | C | 17" x 22" | H-0.75"<br>V-0.50" | Same as Size D and E, Logic Diagrams,<br>Interconnection Diagrams |
| **Full Size<br>or "Bed sheet"** | D | 22" x 34" | H-0.50"<br>V-1.00" | Process Flow Diagrams<br>Piping and Instrumentation Drawings |
|  | E | 34" x 44" | H-1.00"<br>V-0.50" | Orthographic or Scale Drawings<br>Panel Layout and Wiring Same as Size C |

Other sizes, less common, include F thru K and Roll Sizes
(For Architectural Document sizes, see Appendix D.)

*From ANSI/ASME Y14.1*

In some ways, the function of the drawing determines the size. Drawings attached to text specifications are normally "A" size, so they can be easily bound to the parent document. Drawings used for troubleshooting and maintenance will be carried into the field, so it is useful when they easily fold to fit in a pocket. Thus, Loop Diagrams and Installation Details tend to be on "B" or "A" size drawings. Drawings that are scaled or orthographic representations are normally "D" or "E" size since, at typical scales, the area available on the drawing is representative of a reasonable area of the facility.

An owner or design project management establishes the allowable drawing sizes. It may sound trivial, but having a mixture of "C", "D" and "E" size drawings presents a problem for drawing storage files, drawing racks, and even for paper stock in the copiers.

The "C" and "D" size drawings have become somewhat more common with the transition of office work areas from drafting tables to standard desks. A 17" x 22" ("C" size) or 22" x 34" ("D" size) drawing fits nicely on a desktop, yet the available area on the drawing is large enough for practical use. Many non-scaled drawings like P&IDs and Logic Diagrams are often developed in the larger "D" and "E" size format, but they are plotted and distributed as "C" size, or even "B" size, for convenience – not to mention the inherent savings in reproduction costs and space.

Reproduction costs are based on the area of the copy. Smaller is better if the drawing is legible. Orthographic and other drawings produced to scale are normally filed and distributed as full size drawings to maintain their "to scale" nature. Drawings that carry a PE stamp are required by some state laws to be stamped and filed as full size drawings. Today, if reduced sized copies are needed of stamped drawings, they may be made as reduced size photocopies of the wet-stamped original, which can be expensive and time consuming. A limited number of reprographics specialists have the equipment to reduce an "E" size drawing down to "C" size.

There are some projects, and therefore some facilities, that will incorporate multiple copies of "A" or "B" size drawings onto a single "E" size sheet for issue and storage. In our opinion, this should be strongly discouraged. It defeats the key advantages afforded by smaller drawings. Portability and the simplicity of locating a single drawing from a coded drawing number both disappear. One drawing containing four Loop Diagrams will make it hard to relate the loop numbers to the drawing number. Locating the "right" drawing from a list becomes more difficult.

## Non-standard Drawing Sizes

From a practical standpoint, the copy room may only stock three paper sizes – "A", "B" and "E". Drawings have to be made to fit one of those sizes. You may have seen, first hand, difficulties handling non-standard drawing sizes. Have you ever experienced a printer jam while printing documents that came from another country, or possibly a manual downloaded off the Internet? The printer needs instruction on how to handle the unexpected paper size request. European document sizes are different than those in the Americas. ISO metric sized text paper is called A4 here, and it is narrower and taller than 8-1/2" x 11" paper.

Figure 9-2 lists the European and the comparable U.S. sizes. Your printing routine might make the conversion automatically, but the layout of the document may look odd. For the unprepared, working on an international project with offices in Europe and the U.S. can make for some annoying disruptions in work processes in both countries while the conversion tools or procedures are developed.

### Figure 9-2: European Document Sizes

| | ISO Letter | SIZE<br>Width x Length<br>mm (inches) | Like ANSI | ANSI Size<br>Width x Length<br>Inches |
|---|---|---|---|---|
| **Notebook Size** | A4 | 210 x 297<br>(8.3" x 11.7") | A | 8-1/2" x 11" |
| | A3 | 297 x 420<br>(11.7" x 16.5") | B | 11" x 17" |
| **Desktop Size** | A2 | 420 x 594<br>(16.5" x 23.4") | C | 17" x 22" |
| **Full Size<br>or "Bed sheet"** | A1 | 594 x 841<br>(23.4" x 33.1") | D | 22" x 34" |
| | A0 | 841 x 1189<br>(33.1" x 46.8") | E | 34" x 44" |

*From ANSI/ASME Y14.1*

Equipment suppliers are not driven by the same drawing rules as the facility in which the equipment is installed. Consequently, despite all best efforts to maintain site drawing size standards, many different drawing sizes can be found on the vendor drawing stick files. This presents a problem to handle and file. Vendor drawings will range from "A" size installation details to a scale drawing of a paper machine or boiler that can be 15 feet long.

Increasing use of electronic drawings can make drawing storage a bit easier. Drawings are sometimes issued to the purchaser as standard CAD files, assuming the two companies have the same system. Or, more typically, drawings can be exchanged using some universally readable software like Adobe Acrobat™. These are easy to store and read on a desktop computer, but there may be some challenges when a printout is needed. Is the right paper available? Does the third party drawing plot on your system (they don't normally, due to different system setups)?

### Border

The drawing border is, of course, the drawing limit. However, more features can be added that are useful. For starters, the ANSI standard, not to mention years of experience, calls for a margin around the drawing. The margin provides a sacrificial zone where day-to-day damage to the drawing's edges can occur without losing data.

The ANSI margin sizes (see Figure 9-1) seem to be handled in many companies as optional recommendations; many variations on margin size can be experienced. The drawing border itself is a continuous bold rectangle that establishes the fixed outline of the drawing. It is the frame of the picture. The dimensions of the border are also given in Figure 9-1. For drawings printed on a plotter loaded with varying paper roll stock sizes, short light lines can be added in each corner showing where the drawing sheet should be cut. The cut lines are normally about ½" or 1" outside the bold line defining the drawing limit. The cut lines correspond to the edge of the paper, if you pulled it out of a drawer. The plotter paper doesn't normally plot to the edge of the paper, so some definition of the edge is helpful.

A grid system can be added to any drawing to provide a cross-reference, similar to that provided on a road map. For obvious reasons, the grid is probably more useful on larger drawings – "C" size or above. The grid appears between the border and the cut lines. The grid is marked with short hash marks at intervals, every 4-1/4 inches or so. Resulting sections are then identified with sequential numbers and letters on the horizontal and vertical edges. When referring to the location of, say, a control valve on a P&ID, you could use the grid to provide the general area to search for the device, for instance "Grid C-9". The component is quickly located in the vertical C area and the horizontal 9 area.

### Title Block

The title block is the most familiar feature of a drawing template. Almost all title blocks are placed in the lower right corner of a drawing. Figure 9-3 shows a typical Title Block arrangement consisting of the following items:

| Figure 9-3: Typical Title Block | | | | | | |
|---|---|---|---|---|---|---|
| Design By | Date | Owner's Logo | Facility | | | |
| | | Funding No. | Area | | | |
| Drawn By | Date | Designer's Logo | Drawing Type | | | |
| | | Contract No. | | | | |
| Check By | Date | PE Stamp | Drawing Title | | | |
| Appr. By | Date | | Scale | Size Letter | Drawing Number | Rev. |

The drawing number appears in the lower right corner of a drawing, immediately to the left of the drawing revision. Drawing numbers are usually encoded with letters and numerals that define the drawing function. The drawing number may indicate the facility or site, the discipline preparing the drawing, area of the facility to which the drawing applies, type of drawing, and a sequential number. For orthographic drawings, the sequential number may also be encoded with the floor, grid section and elevation. There are as many drawing-numbering schemes as there are faculties that have drawings. However, the better systems encode useful information in the number to make drawing sorting and location easier.

## Drawing Number and Scale

If the drawing types are expected to have multiple sheets, a sheet callout is furnished as part of the template. Logically, the sheet callout appears to the right of the drawing number, and is pre-established in the form: Sheet X of Y. Occasionally, one will see just a sheet callout with no "of" block, but that is rare. It is tidier to know how many sheets are required. The sheet number can be included in the Title Block line itself when multiple sheets occur in a drawing template that has no provision for sheet numbering. We are aware of an example of this. New equipment was added to a P&ID, but there was insufficient room to add it on the existing sheet. P&IDs are supposed to appear in process order, so it was decided to split the "full" P&ID into two sheets, adding the additional equipment into the new space. The title of the P&ID then just had "Sheet 1 of 2" added to the text. Possibly inelegant, but it worked.

The primary scale of the drawing, for orthographic or section drawings, is defined in a dedicated box on the title block. There can be instances on one drawing when a specific area is drawn to a different scale, but that special area scale would be defined in the section or detail title. For P&IDs, Loop Diagrams and other non-scaled drawings, the scale callout should say "None", or some similar statement.

ANSI/ASME Y14.1 calls for identification of the drawing size by adding a block showing the ANSI size letter designation. This practice is becoming less common, but when drawings are filed by size in different flat file drawers, knowing the drawing you are looking for is a "B" size sheet may be critical information so you look in the correct drawer. If your drawings are stored electronically, as many are now, knowing the size may be less important. Furthermore, if your drawing standards are in order and the drawing numbering scheme is effective, adding a separate callout for drawing size becomes superfluous.

The revision is normally provided in the lower right-hand corner of the drawing, although some rely on the revision table alone to carry this information. If the revision callout is provided in the lower right, there should be a table elsewhere on the drawing detailing the revision history. This table is addressed later in the chapter.

### Drawing Titles

Drawing titles are usually multiple lines – sometimes as many as five, but as few as a single line. When multiple lines are used, each line should represent the same information for every drawing within the facility. A typical line convention is provided in Figure 9-3:

- **Facility** – Optional information, the name of the plant site.

- **Area or division** – Commonly provided subset of the plant site or entity furnishing or using the drawings.

- **Drawing Type** – The function or purpose of the drawing. This information may be encoded in the drawing number, but it is helpful to tell those using the drawing its purpose, even when that is self-evident. Here callouts like "Process Flow Diagram", "Piping and Instrumentation Drawings", "Loop Diagram", or "Location Plans" will be found.

- **Drawing Title** – This is really the primary title field of the Title Block. All too often though, some less-than-helpful name makes it into this field. The title should clearly describe the information contained within. For a P&ID, the equipment depicted should be listed, rather than the less-than-helpful "Process Area, sheet 1 of 3", which sometimes happens. The title should explain to those reading a list of drawings exactly what the drawing contains. For a Loop Diagram, the "Service" callout that appears in the Instrument List and on the Specification Forms may be used.

### Owner's Information

Identifying the facility owner is probably called for by the site drawing standards, typically through incorporating a company logo. The size of the logo should balance corporate pride against the value of drawing space. Big logos leave less space for drawing! Many facilities require incorporating funding information in the drawing Title Block – that is, listing the accounting number under which the project will be designed and installed.

### Designer Information

Designer information includes both required information and information the designer would like to add. The single most important "required" information is the small square designed to fit the Professional Engineer's Stamp. State laws are increasingly calling for application of a PE wet stamp – that is, the inked imprint of a Professional Engineer's stamp, signed and dated, on all construction documents, as defined by your state. Shared discipline drawings like a P&ID may carry two or more PE stamps – some combination of the process designer, piping design engineer, and control systems engineer. It should be noted that most states do not allow printing an electronic image of the PE stamp or signature. The "original" drawing has to be physically stamped by the engineer, signed and dated. This may change in the future with the advent of electronic signatures, but don't expect to see the change anytime soon.

The design entity has some pride of ownership of their work producing the drawing, so they may wish to add their corporate logo as well. Some owners set limits on design firm logos by establishing a fixed space within the title block for that purpose. It is very helpful to the design firm to have a space to add their contract number; it makes for simpler tracking of design time and even filing of the drawings.

## Initial Issuing Data

Another common feature of a Title Block is to record the initial issuing data for the drawing – that is, to identify who designed, drew, checked, supervised, and approved the drawing. Dates for each of these steps are recorded in the issuing area of the drawing as well. This information is separate from that contained in the revision block; this area of the title block is reserved for the initial construction release of the drawing. Drawn (drafted), Design, Check and Approve are almost always shown; others are added as called for by project procedures. Due to space limitations, initials are most commonly used. Initials are usually shown and dated by hand for the initial drawing issue. For CAD drawings, subsequent issues will normally record the initials and dates in text, without signatures.

## Other Template Features

Drawings change over time. Information is added and deleted as the process evolves. The evolution of the changes has to be recorded clearly so those using the drawings can quickly see what has changed. Revision blocks are a table used to record the revision activity on a particular drawing. Figure 9-4 shows a Typical Revision Block. ANSI/ASME Y14.1 calls for revision blocks to be located in the upper right corner of the drawing.

The authors are more familiar with revision blocks next to the title block at the bottom of the drawing, but as long as consistency is maintained, either location is fine.

**Figure 9-4: Typical Revision Block**

| 2 | REMOVED T-109 | CAM | 11/25/03 |
|------|---------------|-----|----------|
| 1 | ADDED P-101 | CAM | 01/03/02 |
| REV. | DESCRIPTION | BY | DATE |

In this example, the revision block is located on the bottom of the drawing next to the title block – hence the column callouts. Rev., Description, By and Date are on the bottom of the table, and the revision history proceeds upwards from the bottom. Note that revision 1 is the first entry. The initial issue data was recorded in the title block. (See the previous section on Designer Information.)

There are two ways to handle the situation when there are more revisions than there are blocks available. The revision record can roll over, the oldest revision erased and the newest added as the space is freed; the lines can be re-shuffled (but only if you have a CAD system) or you can just reuse the empty space. Your procedures may call for additional field in the revision block, for instance an "approved" signature block and date.

The description should be as clear as space allows. You can understand why it is less than helpful to those reading the drawing to see the words "General Revision" in the description – they then have to play detective to discover what changed. Complex revisions for some large projects may even cross-reference the description column to another, more detailed text-based revision history document.

## Drawing Change Notices

Sometimes, revisions are issued in small pieces as "Drawing Change Notices" (DCN). These "A" size documents are easy to E-mail and fax, and they allow rapid submittal of drawing revisions without the complexity of issuing a full size drawing. The DCN may show a small area of the drawing with the change incorporated. Usually, text is attached describing the change and listing approval data for the change – that is, approval signatures and change orders. The Revision Block on the drawing may then just list DCNs that were issued earlier, but are now incorporated in the main drawing. The organization to handle a complex system like this is not for every project. However, for complex work involving many different disciplines, this has proven to be an effective system.

## Revision Identification

There is normally insufficient space in the revision block to adequately describe changes to the drawing in text, so some method of locating the changes is needed. Occasionally, the grid marks on the drawing are used to define zones where changes have been made. However, this leaves the challenge of describing the changes in words, and, accordingly, dedicating space on the drawing for that text.

Another, more common, practice is to outline the changes on the drawing itself. Since most lines on a drawing are straight or angular, a distinctive series of connected arcs is used with success. This is called a "revision cloud". The line weight is usually heavier than secondary lines on the drawing but lighter than primary lines; a line width of 0.08" has been used on CAD drawings with good results. See Figure 9-5 for an example of a CAD revision cloud.

### Figure 9-5: Notes and Revision Cloud

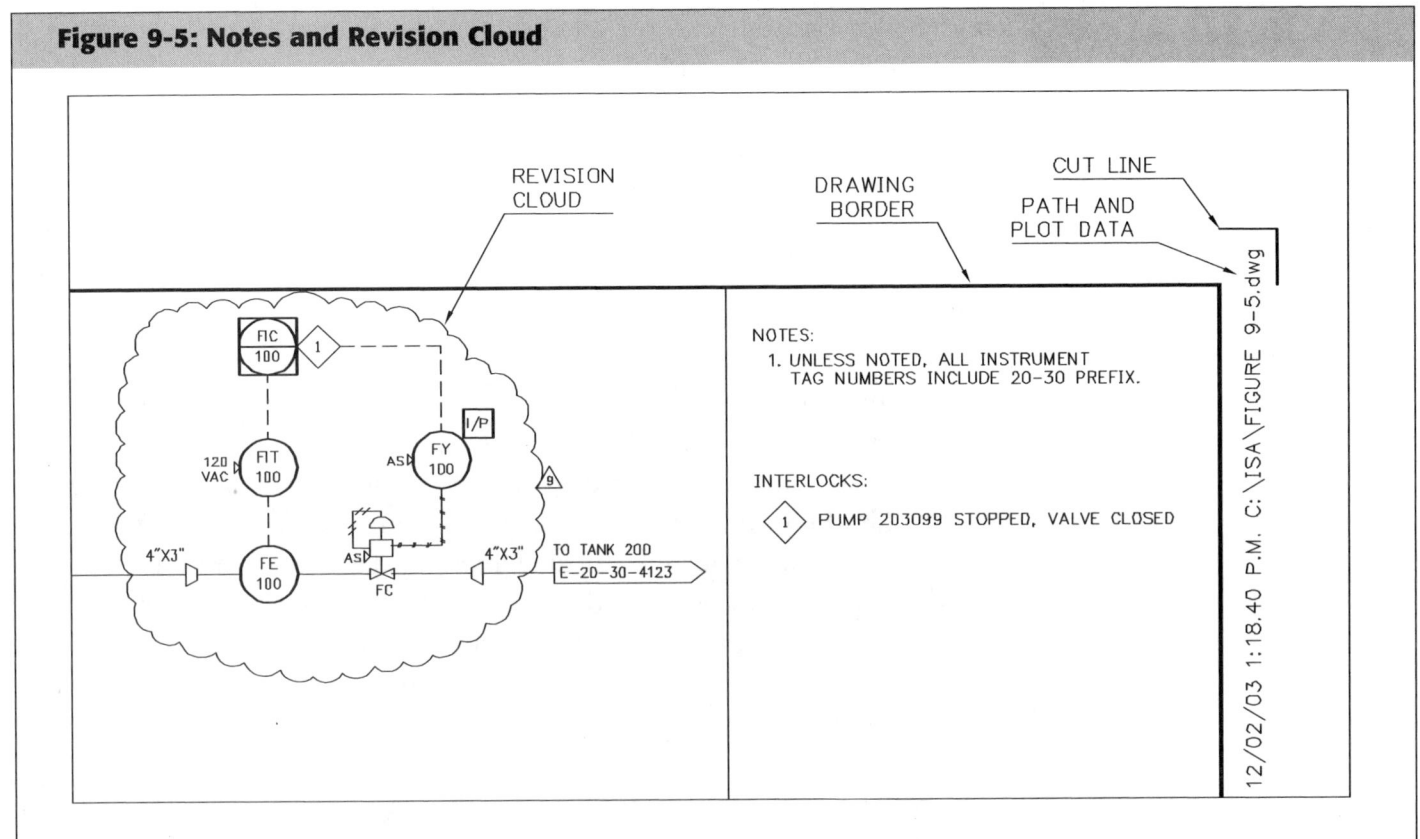

On manual drawings, revision clouds are sometimes drawn with wax pencils that yield a heavy line that can be easily applied and removed. Wax pencils, or grease pencils as they are sometimes known, are available in colors so the clouds are easy to spot. The cloud is keyed to the revision block with the revision number circumscribed by a symbol, typically a triangle. A drawing may have many areas that changed, so there may be many independent revision clouds, each with a revision triangle showing the revision number.

When the drawing is issued again under a new revision, the old clouds are erased and new ones installed outlining the new changes. Sometimes local practice dictates leaving revision triangles behind as a memory jog that something in the area changed in the past. Others find the old triangles to be distracting clutter – and remove them. For manual drawings, the clouds and triangles are normally drawn on the reverse side of the drawing, so they can be removed without disturbing the drawing content.

## Drawing Notes

Drawing notes should, like other features, be placed in the same spot on all drawings for consistency. The notes are numbered sequentially and may be referenced in the body of the drawing. The note section provides space for more lengthy descriptions that won't fit within the body of the drawing. Some companies have success with separate construction notes – that is, notes to the contractor that don't need to exist after installation. These are removed before the drawings are issued for record.

Interlocks, if they are shown on the P&IDs, normally appear above or below the drawing notes, on the right hand side of the sheet. See Figure 9-5.

## References Table

Documents referenced to build the drawing are listed on the references table. The use of this table is somewhat limited for instrumentation and controls. Loop Diagrams should reference the P&ID containing the loop. Location Plans are not normally referenced on a Loop Diagram, since that information is contained in the Instrument Index. However, your procedures may vary. Logic Diagrams and possibly electrical elementary diagrams should reference the P&IDs that show the equipment for which the logic is written. It is sometimes helpful to list a vendor's equipment schematic if that is needed to understand that equipment's operation. Adjacent drawing numbers do not need to be listed for orthographic drawings.

## Plot Data

Some CAD drawings may include a very useful feature of automatically listing the file name and path to the drawing, as well as the plot date, in small text in the margins of the drawing. See Figure 9-5.

When you are madly printing out 200 Loop Diagrams, it is sometimes helpful to know which copy of several is the latest one to print. Also, having the path identified may be useful if you have to find the drawing file later.

## CHAPTER TEN

# Role of Standards and Regulations

According to Webster's Dictionary, a standard is "something established by authority, custom, or general consent as a model or example."[1] The instrumentation and control (I&C) professional must be familiar with many standards. Some are mandatory and must be followed by law. Some standards are optional – they may be followed if the user so chooses. Some standards, after review by a responsible party, may be found to be irrelevant to the activity being preformed and, therefore, may be ignored.

## Mandatory Standards

In the United States, federal, state and local laws establish mandatory requirements to keep everyone safe. These requirements have various names: codes, laws, regulations, requirements, etc. Examples include the Food and Drug Administration Good Manufacturing Practices, National Fire Protection Association (NFPA) Standard 70, National Electric Code (NEC), and Occupational Safety and Health Administration (OSHA) Regulations. There are many others. The U.S. government manages about 50,000 mandatory standards for items as diverse as automobile air bags and missile components.

Local law may even require adhering to past revisions of standards. For instance, most local building authorities require adhering to the National Electric Code (NEC). The NEC is revised every four years. Therefore, local laws have to be revised to reference the latest edition.

Lags sometimes occur. Adopting the revised NEC into local codes takes time, so those of us who do project work normally will check with the local building authority to determine which NEC edition is in effect.

## Food and Drug Administration

The Food and Drug Administration (FDA) issues Good Manufacturing Practices, incorporated in the Quality System Regulations of Section 500 of the Food, Drug and Cosmetic Act. Good Manufacturing Practices require domestic or foreign manufacturers to implement a quality system for their products. Good Manufacturing Practices are in place for such topics as:

- Animal Food and Drugs
- Apple Juice and Cider
- Biologics (vaccines and blood products)

- Cosmetics
- Dietary Supplements
- Food
- Fruits and Vegetables
- Medical Devices
- PET Drugs (Position Emission Tomography)
- Pharmaceuticals for Human Use
- Radiation Emitting Products
- And others

## National Fire Protection Association

The National Fire Protection Association (NFPA) issues many standards. Among them is Standard 70, the National Electric Code (NEC). The NEC defines the way we install safe electrical systems in the United States of America. The standard is huge, covering all electrical installations including house wiring, petroleum plants and even explosives manufacturing. The NEC contains a classification system for electrical construction in areas containing flammable materials. The classification system consists of three parts which, taken together, define the hazard resulting from the flammable materials.

| **Figure 10–1: Hazardous Area Classification** |
|---|
| The National Fire Protection Association (NFPA) issues NFPA 70, the National Electrical Code for areas where flammable materials are present. The National Electrical code defines an area classification system consisting of three parts: |
|     Class |
|     Group |
|     Division |
| **For Example:** Class I Group D, Division 1 |

The first part (Class) denotes the nature of the hazardous material present: flammable gases or vapors, combustible dusts, or ignitable flyings or fibers using Roman numerals I, II, and III. The next part (Group) further specifies the hazardous material groupings using letters A through G – for example, A for acetylene, B for hydrogen, C for carbon monoxide, etc.

There are, of course, many other hazardous materials included under each letter. The third part (Division) defines the risk of the material being present. Division 1 is used where the hazard normally exists. Division 2 is used where the hazard exists in abnormal situations only. The resultant area classification dictates the type of electrical installation necessary.

## Occupational Safety and Health Administration

According to ISA's dictionary, "In 1970 the U.S. Congress passed the Occupational Safety and Health Act which specified the requirements that employers must follow to guard against employees' illness and injury. The Occupational Safety and Health Administration (OSHA) enforces these requirements."[2]

After the Bhopal chemical plant disaster in India, the U.S. Congress ordered OSHA to develop procedures to prevent similar accidents in the U.S. OSHA issued its directives in 1992 as government document 29 CFR 1910.119, *Process safety management of highly hazardous chemicals*. The directive requires that all installations handling certain hazardous materials develop a set of documents, which are defined as follows:

"OSHA requires that employers develop and maintain information about process chemistry, technology, equipment and operating procedures and that this information be communicated to involved employees so that they understand the hazards of the process."[3]

There are three paragraphs in the Process Safety Management directive which list the documents required. Some of the documents require input from the plant instrumentation and controls group.

---

### Figure 10-2: 29 CFR 1910.119 (d) Process Safety Information

#### Typical documents required

- Process flow diagrams
- Process chemical description
- Inventory amounts
- Material safety data sheets (define hazards)
- Safe operating limits
- Consequences of deviation
- Materials of construction
- Electrical classification drawings
- P&IDs
- Relief system design
- Ventilation system design
- Design codes employed
- Material and energy balances for processes built after 05-26-92

---

Paragraph 119 (d), Process Safety Information, lists documents required for process safety. P&IDs are among the listed documents. Of course, the P&IDs should reflect the current state of the plant and should be easily understood by all operating and maintenance personnel. Predictably, this means to the authors that the P&ID symbols and identification methods should be developed and maintained with adherence to ISA-5.1.

**Figure 10-3: 29 CFR 1910 (f) Operating Procedures**

**Documents required**

- Procedures for operation, startup, and shutdown
- Operating limits
- Consequences of deviation
- Procedures for correcting deviation
- Safety and health considerations
- Properties and hazards of chemicals used

- Precautions to prevent exposure
- What to do if exposed
- Safe maintenance procedures
- Quality controls
- Control of inventory

Paragraph 119 (f), Operating Procedures, lists documents required for safe operation. Questions such as the following must be answered:

- How can the plant be operated safely?
- What are the operating limits?
- What happens if the limits are exceeded?
- How can process excursions be resolved?

**Figure 10-4: 29 CFR 1910.119 (l) Management of Change**

- Required for management of change to documentation, process chemicals, technology, equipment and facilities
- Typical documents required
    - Description of change
    - Temporary or permanent
    - Technical basis for change
    - Process hazards
    - Analysis of change
    - Resolutions of recommendations
    - Authorization requirements

Paragraph 119 (l) Management of Change lists documents required to effectively and properly manage a proposed change. Change is defined as anything except replacement in kind. Any proposed change must be defined, analyzed, explained to those concerned, and all questions answered. In other words, the change must be understood. Management must approve proposed changes.

### Industry, Company, Government Entity Standards

Many, if not most, large firms have developed a full bookshelf of practices, specifications and standards that apply to their work. Some government agencies have done the same. Many of these large collections incorporate standards developed by other organizations.

The Department of Energy Architectural Engineering Standard for Instrumentation references several ISA and Institute of Electrical and Electronics Engineers (IEEE) standards. Unless the standards are mandatory, developers of the bookshelf have control of when, if, and how rigorously the bookshelf will be followed. In addition, the developer may change them, although most developers have rigorous revision and approval procedures.

## Consensus Standards

Consensus standards include recommended practices, standards and other documents developed by professional societies and industry organizations. Those standards developed by ISA are the ones most often used by I&C practitioners.

### ISA – The Instrumentation, Systems, and Automation Society

The Instrument Society of America (ISA) was founded in 1945 as a nonprofit educational organization by a small group of dedicated professionals interested in the rapidly emerging field of instrumentation. The society's name has changed, in several steps, to today's "ISA – The Instrumentation, Systems, and Automation Society". The abbreviation ISA has remained constant. The society's intent – to serve as an authoritative technical resource for its members worldwide – has also remained constant.

ISA published its first standard in 1948. Since then ISA has developed more than 150 automation, instrumentation, and control system Standards, Recommended Practices and Technical Reports. The differences in these documents are defined by ISA as follows:

"*Standard*: A document that embodies requirements (normative material) that if not followed, could directly affect safety, interchangeability, performance, or test results…."

"*Recommended Practice*: A document that embodies recommendations (informative material) that are likely to change because of technological progress or user experience, or which must often be modified in use to accommodate specific needs or problems of the user of the document."

"*Technical Report*: A document that embodies informative material."[4]

---

**OSHA Compliance Challenging**

The OSHA regulations sound complete and complex, and some plants achieve compliance only reluctantly. An older plant I once visited included two areas that handled hazardous materials and several areas that did not. Some of the original information and documentation was no longer available or useful.

For example, the plant P&IDs had been developed over a number of years, as process areas were added to the plant. Different engineering contractors designed the areas, each using their own standards. Different specialists had revised P&IDs over the years. There was no firm plant standard, so revisions did not always agree. Therefore, developing the whole documentation package was not easy.

However, after successfully developing the OSHA documentation package for the required plant areas, plant management decided it was a valuable exercise. They proceeded to develop the whole documentation package for the rest of the plant, and determined it was very worthwhile.

### To clarify further . . .

Perhaps a bit of clarification of these definitions is in order. "Normative" means normal. For the word "normative", substitute "normal practice". For the word "informative", substitute "giving information". Then the three documents might be defined as follows:

*Standard:* Guidelines which follow normal practice on a specific subject. The guidelines are unlikely to change. A large number of unbiased experts have reviewed the document and agree the stated guidelines are very important to instrumentation and control personnel.

*Recommended Practice:* Information guidelines on a specific subject based on the best information available today. These guidelines are subject to change. A large number of experts have reviewed the document and have agreed that the guidelines are very important.

*Technical Report:* Information on a specific subject. A number of experts have reviewed the document and have agreed it is of interest.

ISA follows formal rigorous procedures to develop Standards, Recommended Practices and Technical Reports. These procedures assure the documents produced are accurate, balanced, and fair. About half of ISA Standards, Recommended Practices, and Technical Reports have also received American National Standards Institute (ANSI) approval.

ANSI was established in 1918 as the American Engineering Standards Committee. Its purpose is to provide standards to improve the quality and methods of American industry. ANSI approval procedures are different, but equally or more rigorous than that of the ISA. ANSI has accredited ISA to develop ANSI/ISA documents.

---

**Figure 10-5: ISA-5 Documentation Series**

- ISA-5.1 Instrumentation Symbols and Identification
- ISA-5.2 Binary Logic Diagrams for Process Operations
- ISA-5.3 Graphic Symbols for Distributed Control/Shared Display Instrumentation, Logic and Computer Systems
- ISA-5.4 Instrument Loop Diagrams
- ISA-5.5 Graphic Symbols for Process Displays

---

Many quotes from the ISA-5 Documentation Series are included in the previous chapters. The series includes the following:
  - *ISA-5.1-1984-(R1992), Instrumentation Symbols and Identification,* presents a comprehensive set of symbols and the identification methods

used to designate instrumentation and control devices on design documents. ISA-5.1 is the standard most often used throughout the world for identifying instrumentation and control devices and systems on P&IDs.

- **ISA-5.2-1976-(R1992), *Binary Logic Diagrams for Process Operations*,** provides the additional symbols used on Logic Diagrams and describes a method of logic diagramming for on-off or binary control systems.

- **ISA-5.3-1983, *Graphic Symbols for Distributed Control/Shared Display Instrumentation, Logic and Computer Systems*,** developed to supplement ISA-5.1 for identification of instruments and control devices when shared control shared display systems were used. However, the key elements of ISA-5.3 are now included in ISA-5.1, and ISA-5.3 will be withdrawn.

- **ISA-5.4-1991, *Instrument Loop Diagrams*,** includes some additional symbols and six typical instrument Loop Diagrams – two each for pneumatic, electronic, and shared display and control loops. One of each type includes the minimum requirements and the second includes the minimum plus optional items.

- **ISA-5.5-1985, *Graphic Symbols for Process Displays*,** establishes a set of symbols used in process displays to depict processes and process equipment.

- Other ISA Standards, which are useful in developing a design package, include:

- **ISA-20-1981, *Specification Forms for Process Measurement and Control Instruments, Primary Elements and Control Valves*,** includes specification forms and instructions for use in defining many instrumentation and control devices.

- **ISA TR20.00.01-2001, *Specification Forms for Process Measurement and Control Instruments, Part 1: General Considerations*** is an updated version of ISA 20, which includes 50 totally new and technically updated specification forms. The TR means Technical Report, a designation that will allow frequent revisions and updates without the protracted approval process associated with standards. The TR is a work in progress. For the latest information on specification forms see the ISA website, **http://www.isa.org**.

- **ANSI/ISA-84.01-1996, *Application of Safety Instrumented Systems for the Process Industries*.** The title of this standard explains its contents. Its intent is to define requirements for designing, manufacturing, installing, commissioning, testing, maintaining, and operating of safety instrumented systems (SIS).

- **ANSI/ISA-88.01-1995, *Batch Control Part 1: Models and Terminology*,** which shows relationships between batch control-related models and terminology.

ISA Standards, Recommended Practices, and Technical Reports cover more than 40 different subjects. Some of the most numerous documents cover subjects such as air measurement transducers, control valves and actuators, documentation, electrical equipment for hazardous locations, fossil power plants, measurement transducers, nuclear power plants, signal compatibility and temperature.

## Other Consensus Standards

In addition to ISA, other organizations also develop documents to guide the professional. These organizations include American Petroleum Institute (API), American Society of Mechanical Engineers (ASME), National Electrical Manufacturers Association (NEMA), Process Industry Practices (PIP), Scientific Apparatus Manufacturers Association (SAMA), and Technical Association of the Pulp and Paper Industry (TAPPI), among others.

## American Petroleum Institute (API)

API is the primary oil and gas industry trade association with a membership of about 400 organizations from all parts of the industry. API aims to provide information and aid to the oil and gas industry to improve efficiency and cost effectiveness, comply with governmental rules and regulations, safeguard health, improve safety, and protect the environment. API develops and distributes technical standards and other publications on oil industry subjects such as:

| | |
|---|---|
| Exploration and Production | Safety and Fire Protection |
| Petroleum Measurement | Storage Tanks |
| Marine Transportation | Valves |
| Marketing | Industry Training |
| Pipeline Transportation | Health, Environment & Safety |
| Refinery | Instrument Installation |

## American Society of Mechanical Engineers (ASME)

There are several ASME standards that can be useful when developing instrumentation and control documents.

*Y32.11 Graphical Symbols for Process Flow Diagram*, includes a series of symbols that depict:

| | |
|---|---|
| Lines | Pressure Vessels |
| Valves | Dryers |
| Furnaces and Boilers | Material Handling Equipment |
| Heat Transfer Equipment | Size Reduction Equipment |
| Pumps and Compressors | Processing Equipment |
| Drivers | Separators |
| Storage Vessels | |

*ASA Z32.2.3* (re-designated **ANSI/ASME Y32.2.3**, *Graphic Symbols for Pipe Fittings, Valves, and Piping*), is a guide for depicting pipe, fittings, and manually operated valves.

## National Electrical Manufacturers Association (NEMA)

*NEMA Standards Publications No. 250-2003 Enclosures for Electrical Equipment (1000 Volts Maximum)*, covers the classification and description of enclosures for electrical equipment and the environmental conditions each one can withstand. This includes information on NEMA enclosures, types 1 through 13.

## Process Industry Practices (PIP)

PIP is a coalition of process industry owners and engineering/construction contractors that has developed and is continuing to develop recommended practices for all areas of the engineering, procurement, and construction process. ISA and PIP have entered into a cooperative agreement that allows ISA to use the content of **PIP PIC001, *Piping and Instrumentation Diagram Documentation Criteria***. This PIP work is currently available as an ISA draft standard for trial use, **ISA-DSTU-5.07.01-2002, *Piping and Instrumentation Diagram Documentation Criteria***. It is expected that, following an ISA trial-use period, this draft standard – revised as necessary – will be submitted to ANSI for approval as an American National Standard.

PIP's Process Control Practices include the following subjects:

| | | |
|---|---|---|
| Control Regulators | Flow | Process Analyzers |
| Control Valves | General | Pressure |
| Documentation | Instrument Air | Package Systems |
| Differential Pressure | Instrument Piping | Safety Systems |
| Electrical | Level Instrumentation | Temperature |
| | | Weighing Systems |

For further information on PIP, see their web site, **http://www.pip.org/**.

## Scientific Apparatus Makers Association, (SAMA)

SAMA no longer exists as an organization. Some years ago, SAMA developed **PMC 22.1, *Functional Diagramming of Instrument and Control Systems***. PMC 22.1 is still in use in some industries, especially to document boiler control systems and functions. It details the function of the control system by a series of symbols, showing, for example, the modes of control in a controller, the proportional (gain), integral (reset), and derivative (rate). Functional Diagrams supplement, but do not replace, P&IDs.

At the time of their withdrawal, SAMA gave ISA permission to use the SAMA work. Portions of PMC 22.1 are directly incorporated into the current edition of ISA-5.1. The entire PMC 22.1 is in the process of being issued by an ISA standards committee.

## Technical Association of the Pulp and Paper Industry (TAPPI)

TAPPI is a technical association founded in 1915. Over 20,000 individual members are involved in the pulp, paper, and converting industry. TAPPI has developed and revised Test Methods and Standard Practices for use in the industry.

## Location-Specific Documents

These are the standards and other documents developed for a specific location or a specific project. They may be a simple document or they may be a very specific and detailed set of books. They may be as basic as the legend sheet that should (but is not always) included with every issue of a set of P&IDs. The legend sheet, or sheets, should indicate the symbols, abbreviations, and letter designations used in the attached documents. Some projects add additional information including typical P&IDs and other sample documents.

There are two international standards organizations that develop standards for worldwide use — the International Electrotechnical Commission (IEC) and the International Organization for Standardization (ISO).

## International Electrotechnical Commission

The International Electrotechnical Commission (IEC) (**http://www.iec.ch**) is a standards organization dealing with electrical, electronic, and related technologies. Many of its standards are developed jointly with ISO.

IEC is made up a representatives of national standards bodies. It was founded in 1906 and currently has more than 60 participating countries. IEC headquarters are in Geneva, Switzerland. The United States member of IEC is American National Standards Institute (ANSI).

## International Organization for Standardization (ISO)

ISO (**http://www.iso.ch**) is unique because it is composed of the national standards institutes of 147 countries. Some of these standards institutes are part of their country's government, while others are in their country's private sector. ISO develops voluntary standards that are very useful in facilitating international trade. For example, the ISO 9000 series are generic quality management standards, which confirm that a location has a system in place to maintain quality by managing its processes or activities. The procedures to confirm conformity to the standard have been developed by ISO together with the International Electrotechnical Commission (IEC). The conformity assessment, or outside audit, is carried out by other organizations.

1. *Webster's Seventh New Collegiate Dictionary*, (Springfield, MA, G&C Merriam Co., 1967) p.853

2. op cit *The Automation, Systems, and Instrumentation Dictionary*, 4th edition (Research Triangle Park, NC: ISA — The Instrumentation, Systems, and Automation Society, 1984), p354.

3. ISA-FG-15, Developing and Applying Standard Instrumentation & Control Documentation, Version 2.2 Slide 30

4. ISA Standards and Practices Department Accredited Procedures (Research Triangle Park, NC, ISA, January 31, 2001) p.5

# Appendix A
# Answers to Chapter 2 Exercise

**Figure 2-19, Descriptions**

**Figure 2-20, Symbols**

| | | |
|---|---|---|
| 1. U | 8. H | 15. L or O |
| 2. A | 9. S | 16. N |
| 3. C | 10. F | 17. Q |
| 4. P | 11. B | 18. R |
| 5. T | 12. L or O | 19. K |
| 6. D | 13. E | 20. M |
| 7. J | 14. G | 21. I |

# Appendix B
# Abbreviations

| Abbreviation | Explanation |
|---|---|
| ac | Alternating Current |
| AI | Analog Input |
| ANSI | American National Standards Institute |
| AO | Analog Output |
| API | American Petroleum Institute |
| A/R | As Required |
| ARO | After Receipt of Order |
| ASA | American Standards Association (NOTE: Superceded by ANSI) |
| ASME | American Society of Mechanical Engineers |
| BPCS | Basic Process Control System |
| CAD | Computer-Aided Design or Drafting |
| CADD | Computer-Aided Design and Drafting |
| CE | European Community Mark, indicating compliance with European requirements |
| CS | Carbon Steel |
| CSA | Canadian Standards Association |
| dc | Direct Current |
| DCN | Drawing Change Notice |
| DCS | Distributed Control System |
| DI | Digital Input |
| DIR | Direct |
| DO | Digital Output |
| dp | Differential Pressure |
| EPC | Engineering - Procurement - Construction |
| F | Fahrenheit |
| FC | Fail Closed |
| FDA | Food and Drug Administration |
| FI | Fail Indeterminate |
| FL | Fail Last, or Locked |
| FM | Factory Mutual |
| FO | Fail Open |
| FOB | Freight On Board |
| I&C | Instrumentation and Control |
| I/O | Input/Output |

| Abbreviation | Explanation |
|---|---|
| I/P | Current to Pneumatic (converter) |
| IA | Instrument Air |
| IEC | International Electrotechnical Commission |
| IEEE | Institute of Electrical and Electronics Engineers |
| IL | Instrument List |
| ISA | ISA — The Instrumentation, Systems, and Automation Society |
| ISA Dictionary | *The Automation, Systems, and Instrumentation Dictionary*, 4th edition |
| ISA-FG15 | ISA Training Course FG15, Developing amd Applying Standard Instrumentation & Control Documentation, Version 2.2 |
| ISA-S106 | ISA Training Course S106, Version 2.0, Introduction to Measurement and Control — Technology, Industries & People |
| ISO | International Organization for Standardization |
| KO | Knock Out |
| mA | Milliamperes |
| MW | Manway or Access Port |
| NEC | National Electric Code |
| NEMA | National Electric Manufacturers Association |
| NFPA | National Fire Protection Association |
| OSHA | Occupational Safety and Health Administration |
| OWS | Oily Water Sewer |
| PAS | Plant Automation System |
| P&ID | Piping and Instrumentation Drawing |
| PE | Professional Engineer (licensed) |
| PES | Programmable Electronic System |
| PFD | Process Flow Diagram |
| PID | Proportional-Integral-Derivative |
| PIP | Process Industries Practices, founded by Construction Industries Institute (CII) |
| PLC | Programmable Logic Controller |
| PP | Personnel Protection |
| PPA | Prepay and Add (freight costs) |
| psi | Pounds per square inch |
| psig | Pounds per square inch gauge |
| PV | Process Variable |
| REV | Reverse |
| RFQ | Request For Quotation |
| ROI | Return On Investment |
| SAMA | Scientific Apparatus Makers Association |
| SCADA | Supervisory Control and Date Acquisition |
| SIS | Safety Instrumented Systems |
| SLDC | Single Loop Digital Controller |
| SP | Set Point |
| TAPPI | Technical Association of the Pulp and Paper Industry |
| T&C | Terms and Conditions |
| UL | Underwriters Laboratory |
| v ac | Volts - alternating current |
| v dc | Volts - direct current |

# Appendix C
# Typical ISA-TR20.00.01
# Specification Form

The following page was excerpted from the CD of ISA-TR20.00.01-2001, Specification Forms for Process Measurement and Control Instruments, Part 1: General Considerations. It is the first page of the device specification for a displacer-type level transmitter or local controller, Form 20L2121 Rev 0. The CD also provides additional pages that form a "pick list" (not included in this book). The "pick list" offers a choice for each of the lines of the device specification. For example, for line 13, "process connection, nominal size and rating", it is possible to select any one of eight sizes ranging from 1½ inches to 8 inches, or 50 mm to 100 mm for the process connection size.

| # | | | | # | |
|---|---|---|---|---|---|
| 1 | RESPONSIBLE ORGANIZATION | DISPLACER-TYPE LEVEL TRANSMITTER | | 6 | SPECIFICATION IDENTIFICATIONS |
| 2 | | OR LOCAL CONTROLLER | | 7 | Document no |
| 3 | (ISA) | Device Specification | | 8 | Latest revision        Date |
| 4 | | | | 9 | Issue status |
| 5 | | | | 10 | |

| # | BODY OR CAGE | | # | TRANSMITTER OR CONTROLLER Continued |
|---|---|---|---|---|
| 11 | BODY OR CAGE | | 60 | TRANSMITTER OR CONTROLLER Continued |
| 12 | Body/Cage type | | 61 | Cert/Approval type |
| 13 | Process conn nominal size | Rating | 62 | Mounting location/type |
| 14 | Process conn termn type | Style | 63 | Span-Zero adjust lct |
| 15 | Flange facing finish | | 64 | Enclosure material |
| 16 | Lower conn location | Upper | 65 | |
| 17 | Gage glass conn nom size | Rating | 66 | |
| 18 | Gage glass conn type | Style | 67 | |
| 19 | Vent/Drain conn nom size | Rating | 68 | PERFORMANCE CHARACTERISTICS |
| 20 | Vent/Drain termn type | Style | 69 | Max press at design temp        At |
| 21 | Head orientation/type | | 70 | Min working temperature        Max |
| 22 | Extension/Heat insulator | | 71 | Accuracy rating |
| 23 | Body/Cage material | | 72 | Level Lower Range-Limit        URL |
| 24 | Bolting material | | 73 | Min level span        Max |
| 25 | Gasket/O ring material | | 74 | Min differential sp gr |
| 26 | | | 75 | Min ambient working temp        Max |
| 27 | | | 76 | Contacts ac rating        At max |
| 28 | | | 77 | Contacts dc rating        At max |
| 29 | SENSING ELEMENT | | 78 | |
| 30 | Sensor type | | 79 | |
| 31 | Sp gr Lower Range-Limit | URL | 80 | |
| 32 | Displacer diameter | Length | 81 | ACCESSORIES |
| 33 | Extension length | | 82 | Connecting cables length |
| 34 | Displacer material | | 83 | Air set filter style |
| 35 | Torque tube/Spring matl | | 84 | Air set gauges |
| 36 | Extension/Cable material | | 85 | Gage glass style |
| 37 | Trim material | | 86 | Gage valve style |
| 38 | | | 87 | |
| 39 | | | 88 | |
| 40 | CONNECTION HEAD | | 89 | SPECIAL REQUIREMENTS |
| 41 | Type | | 90 | Custom tag |
| 42 | Enclosure type no/class | | 91 | Reference specification |
| 43 | Signal termination type | | 92 | Special preparation |
| 44 | Cert/Approval type | | 93 | Compliance standard |
| 45 | Enclosure material | | 94 | Construction code |
| 46 | | | 95 | Software configuration |
| 47 | | | 96 | |
| 48 | | | 97 | |
| 49 | TRANSMITTER OR CONTROLLER w/wo SWITCHES | | 98 | PHYSICAL DATA |
| 50 | Housing type | | 99 | Estimated weight |
| 51 | Output signal type | | 100 | Overall height |
| 52 | Enclosure type no/class | | 101 | Removal clearance |
| 53 | Control mode | | 102 | Upper to lower conn lg |
| 54 | Digital communication std | | 103 | Lower to drain conn lg |
| 55 | Signal power source | | 104 | Signal conn nominal size        Style |
| 56 | Contacts arrangement | Quantity | 105 | Mfr reference dwg |
| 57 | Transient protection | | 106 | |
| 58 | Integral indicator style | | 107 | |
| 59 | Signal termination type | | 108 | |

| 110 | CALIBRATIONS AND TEST | | INPUT OR SETPOINT | | | OUTPUT OR SCALE | |
|---|---|---|---|---|---|---|---|
| 111 | TAG NO/FUNCTIONAL IDENT | MEAS/SIGNAL/TEST | LRV | URV | ACTION | LRV | URV |
| 112 | | Level-Analog output | | | | | |
| 113 | | Level-Digital output | | | | | |
| 114 | | Level-Scale | | | | | |
| 115 | | Level setpoint 1-Output | | | | | |
| 116 | | Level setpoint 2-Output | | | | | |
| 117 | | Test pressure | | | | | |

| 118 | COMPONENT IDENTIFICATIONS | | |
|---|---|---|---|
| 119 | COMPONENT TYPE | MANUFACTURER | MODEL NUMBER |
| 120 | | | |
| 121 | | | |
| 122 | | | |
| 123 | | | |
| 124 | | | |
| 125 | | | |

| Rev | Date | Revision Description | By | Appv1 | Appv2 | Appv3 | REMARKS |
|---|---|---|---|---|---|---|---|
| | | | | | | | |
| | | | | | | | |
| | | | | | | | |

Form: 20L2121 Rev 0

# Appendix D
# Drawing Sizes

The tables below compare architectural and engineering drawing sizes. Until 1995, sizes A through E, below, were used in the United States for **both** architectural and engineering drawings.

| Engineering Drawing Sizes | | |
|---|---|---|
| **ANSI Size** | **Dimensions (inches)** | **Dimensions (millimeters)** |
| A | 8.5 × 11 | 216 × 279 |
| B | 11 × 17 | 279 × 432 |
| C | 17 × 22 | 432 × 559 |
| D | 22 × 34 | 559 × 864 |
| E | 34 × 44 | 864 × 1,118 |

Since 1995, the sizes ARCH 1 through 6 below have been used for Architectural Drawings.

| Architectural Paper Sizes | | |
|---|---|---|
| **Size** | **Dimensions (inches)** | **Dimensions (millimeters)** |
| ARCH 1 | 9 × 12 | 229 × 305 |
| ARCH 2 | 12 × 18 | 305 × 457 |
| ARCH 3 | 18 × 24 | 457 × 610 |
| ARCH 4 | 24 × 36 | 610 × 914 |
| ARCH 5 | 30 × 42 | 762 × 1,067 |
| ARCH 6 | 36 × 48 | 914 × 1,219 |

# Appendix E
# Recommended References

The authors asked several I&CS practitioners what books they have near their desk for ready reference. We also have added our ideas to produce the following list.

## Control Valves

- *Control Valves*, Guy Borden, Jr., Editor, an ISA publication
  An overall view of control valves – their application, construction, design, and maintenance. The information assists engineers during control valve selection and operation.

- *Control Valve Primer, a User's Guide*, H. D. Baumann, an ISA publication
  This book gives practical advice on how to apply control valves. It presents insights on valve sizing, smart valve positioners, field-based architecture, network system technology, and control loop performance evaluation.

- *The Control Valve Handbook*, Fisher Controls, 2nd edition, third printing
  There are several good manufacturers' handbooks. This one happens to be the one with which we are most familiar. It is a good reference that explains the function, features, terminology and individual components that comprise a control valve. The handbook explains how a control valve works and may provide insight into why your valve is behaving oddly. Even old timers will revisit the handbook from time to time.

## Electrical

- National Fire Protection Association (NFPA) Standard 70, National Electric Code (NEC) (latest edition)
  Well, you can't, or should not, play the game without knowing the rules, and here are the rules, electrically anyway. There are good classes out there that will introduce neophytes to "the code", as well as offer refresher courses to the rest of us.

## Flow

- *Flow of Fluids through Valves, Fittings and Pipe*, Crane Technical Paper 410, 21st Printing, 1982, The Crane Company

   This is that little orange spiral bound book seen on control systems engineers' desks, as well as on the desks of process and mechanical engineers. Crane Technical Paper 410 contains an invaluable memory jogging multi-page listing of fluid flow equations, an explanation of how to calculate fluid flow losses through valves and fittings, fluid property tables, and even tables showing conversion factors. It is probably the most used reference book on many desks.

- *Flow Measurement Engineering Handbook*, 2nd edition, R.W. Miller, McGraw Hill

   This is a fine handbook providing clear, concise, and, if you want it, detailed and technical information about the science and application of flow measurement devices. It is a good source for orifice plate bore calculations. We recommend that everyone run through the manual calculation occasionally so you can better appreciate the software. It also is good to remember the factors that influence the reading you see after the system is installed.

- *Flow Measurement*, 2nd edition, D. W. Spitzer, Editor, an ISA publication

   An excellent handbook covering the whole array of flow meters. Emphasis is placed on the accuracy of measurement and how to ensure it. The book includes topics on forecasting flow conditions, flow meter selection, installation, calibration, and maintenance.

## Standards

- ISA-5.1-1984-(R1992), Instrument Symbols and Identification

   This is the oracle for anyone documenting a control system. From P&IDs to sketches, this is the standard for the symbolism used. This was the source document for thousands of P&ID legend sheets – and hundreds of thousands of spirited discussions known colloquially as arguments.

- ISA-5.4-1991, Instrument Loop Diagrams

   So what makes a Loop Diagram? Well, this standard not only establishes the content of the ubiquitous loop, but it also presents increasing levels of complexity that can be seen in typical Loop Diagrams. ISA-5.4 includes sample loop diagrams for pneumatic, electronic and shared display systems. The sample loops define the minimum requirements for Loop Diagrams as well as additional optional items that can be included.

- ISA-20-1981, Specification Forms for Process Measurement and Control Instruments, Primary Elements, and Control Valves

   Not only are there blank specification data sheets in this standard, but next to each data sheet template, a line-by-line listing can be found telling the user what information each block or datum was supposed to provide.

## Temperature

- *The Temperature Handbook*, 27th edition (or latest), Omega Technologies

  Almost everyone has at least one Omega handbook at their desk. The one referred to most often might be *The Temperature Handbook*, as it includes so much information about thermocouples, RTDs and other temperature elements, transmitters and controllers – "hey, what color code is used for a type K thermocouple, anyway?". Contains easy-to-understand technical information on many topics.

## Vendors

- **Vendor catalogs, CDs, or more commonly, Web access to catalog information.**

  If you are working with instrumentation, Internet access to manufacturer's current specification information and model numbers is essentially a requirement. Some manufacturers have well organized, intuitive and accessible Web sites; some have simply awful Web sites. Thankfully, the better instrument and control valve suppliers have user-friendly access to ALL their technical data, which we feel is a critical criterion for instrumentation and controls professionals. If your supplier doesn't have accessible, current and complete information available on the Web, including installation and termination information, ask them to fix the site or think about finding another supplier. Hopefully, all our suppliers will eventually respond to suggestions, so it is up to us to share the pain when the sites don't work like we need them to work. The worst offenders offer minimal, glossy information with no technical "meat", telling us to contact our local representative for details. In our experience, it is better to simply use a different supplier. Many instrument users link to vendor Web sites or have loaded the manufacturers' installation, operation and maintenance manuals onto their Intranet, so there are no libraries of out-of-date information to lead us astray.

- *ISA Directory*, **Latest Edition**

  The *ISA Directory* is the ubiquitous desk reference for locating and contacting instrument suppliers. It is invaluable, particularly when you have to depart from your "comfort zone" of standard devices. When you have to come up with a widget that you have never specified before, this directory gives you names to call.

# Index